MW00443917

Global Brooklyn

Also Available from Bloomsbury

Bite Me, *Fabio Parasecoli*
The Bloomsbury Handbook of Food and Popular Culture,
edited by Kathleen LeBesco and Peter Naccarato
Digital Food, *Tania Lewis*

Global Brooklyn

Designing Food Experiences in World Cities

Edited by
Fabio Parasecoli and Mateusz Halawa

BLOOMSBURY ACADEMIC
LONDON • NEW YORK • OXFORD • NEW DELHI • SYDNEY

BLOOMSBURY ACADEMIC
Bloomsbury Publishing Plc
50 Bedford Square, London, WC1B 3DP, UK
1385 Broadway, New York, NY 10018, USA

BLOOMSBURY, BLOOMSBURY ACADEMIC and the Diana logo are trademarks
of Bloomsbury Publishing Plc

First published in Great Britain 2021

Copyright © Fabio Parasecoli, Mateusz Halawa, and contributors, 2021

Fabio Parasecoli, Mateusz Halawa, and contributors have asserted their right under the
Copyright, Designs and Patents Act, 1988, to be identified as authors of this work.

For legal purposes the Acknowledgments on pp. vii–x constitute an extension of
this copyright page.

Cover design by Dani Leigh Design
Cover photos: bharath g s, Tomas Jasovsky, Ahmad Barshod and bantersnaps on Unsplash

All rights reserved. No part of this publication may be reproduced or transmitted in
any form or by any means, electronic or mechanical, including photocopying, recording,
or any information storage or retrieval system, without prior permission in writing
from the publishers.

Bloomsbury Publishing Plc does not have any control over, or responsibility for, any
third-party websites referred to or in this book. All internet addresses given in this
book were correct at the time of going to press. The author and publisher regret any
inconvenience caused if addresses have changed or sites have ceased to exist, but can
accept no responsibility for any such changes.

A catalogue record for this book is available from the British Library.

A catalog record for this book is available from the Library of Congress.

ISBN: HB: 978-1-3501-4447-7
PB: 978-1-3501-4446-0
ePDF: 978-1-3501-4449-1
eBook: 978-1-3501-4448-4

Typeset by Deanta Global Publishing Services, Chennai, India
Printed and bound in Great Britain

To find out more about our authors and books visit www.bloomsbury.com and
sign up for our newsletters.

Contents

Acknowledgments

We want to thank the reviewers for the *Food, Culture and Society* article that initiated the project that led to this volume, as well as the anonymous readers who evaluated the proposal for this book, for their constructive observations and their appreciation of the direction in which we were taking our research.

Our gratitude goes to the Scuola Politecnica di Design, the University of Copenhagen, the Skylab FoodLab at the Technical University of Denmark, the Association for the Study of Food and Society, the University of Gastronomic Sciences, and New York University, where we presented our work in progress, receiving invaluable feedback.

We would also like to thank Miriam Cantwell, who spearheaded this project at Bloomsbury before moving to a new phase of her career, and Lily McMahon who supported us in completing the work.

Thanks to the colleagues and administration at the Institute of Philosophy and Sociology of the Polish Academy of Sciences in Warsaw, especially Marcin Serafin and Marta Olcoń-Kubicka at the Max Planck Partner Group for the Sociology of Economic Life, for their sustained and stimulating comments and questions that made this work better, and to Agata Bachórz, who brought to our work an expert and imaginative knowledge of food practices in Poland.

This work was supported by the National Science Centre, Poland (grant number DEC-2017/27/B/HS2/01338) and by a start-up fund provided by New York University, Steinhardt.

Parts of the introduction and the conclusion were previously published in the article: Halawa, Mateusz, and Fabio Parasecoli. 2019. "Eating and Drinking in Global Brooklyn." *Food, Culture and Society* 22 (4): 387–406.

* * *

I want to express my sincere appreciation to my coauthor Mateusz Halawa, who fatefully came up with the initial concept of "Global Brooklyn" back in 2016, just as I was starting to explore the Polish culinary landscape and he offered to help me navigate the new environment. We met as fellows in the first cohort of the Graduate Institute for Design, Ethnography, and Social Thought at The

New School, where we realized that we were able to communicate creatively across very different fields (economic sociology and food studies) through our shared interest in materials, spaces, and their effect on people. Those exchanges became the original inspiration for this project. Mateusz, this book would not exist without your brilliance and critical acumen. Thank you for your patience with my restlessness and idiosyncrasies. I have learned a lot from you, and not only in terms of literatures I was not familiar with. Writing and editing together is a grueling process. We made it out in one piece (although with a few bruises). These pages are a testament to the fruitfulness and effectiveness of our collaboration.

I want to thank my family and friends, wherever they may be in the world (distance does not matter). They have always been there to support me and my work as a scholar, researcher, and author. Special thanks to Doran Ricks: we have been through thick and thin, and I know I can always rely on you.

I am deeply grateful to all who sometimes gently and joyfully, sometimes critically and bluntly, shared observations, reflections, and ideas with me, contributing to the development of this project. Besides those already mentioned above, I want to acknowledge the students and colleagues from the Nutrition and Food Studies Department at NYU, NYU Shanghai, The New School, the Association for the Study of Food and Society, Food Design North America, the Red Latinoamericana de Food Design, the Bologna Business School, the University of Gastronomic Studies in Pollenzo, Elisava School of Design and Engineering in Barcelona.

In particular, I want to thank those who have explored with me the topics of design and food design, in their complex connections with sustainability, justice, and politics: in no specific order, Adam Brent, Joel Towers, Susan Yelavich, Bea Banu, Stefani Bardin, Emilie Baltz, Jehangir Mehta, Sonia Massari, Christy Spackman, Antonello Fusetti, Davide Chiesa, Stefano Maffei, Franco Fassio, Sonja Stummerer and Martin Hablesreiter (aka Honey and Bunny), Catherine Flood, May Rosenthal Sloan, Carolien Niebling, Rositsa Ilieva, Karen Karp, Pedro Reissig, Freddy Zapata, Juan José Arango Correa, Monika Kucia, Karol Murlak, Daniel Bergara, Nataly Restrepo, Peter Kim, Andrés Sicard Currea, Marije Vogelzang, Katja Grujters, Mariana Eidler Diaz, Albert Fuster, Paolo Sustersic, Armin Zadakbar, Marti Guixé, Malena Pasin, Damian Valles, Monica Silva, Gabriela Velarezo, Noel Gonzales, Ricardo Yudi, Paolo Barichella, Marc Bretillot, Ido Garini, Rick Schifferstein, Francesca Zampollo, Ricardo Bonacho, Fernando Laposse, Massimiliano Tonelli, Federico De Cesare Viola, Fernando

Moreira da Silva, Charlotte Biltekoff, Richard Mitchell, Sara Roversi, Roberto Flore, Joaquim Fernandes Nieto, Anna Greenspan, Gabriele Tempesta, Julian Liu, Wei Shaonong.

—*Fabio Parasecoli*

* * *

I want to thank my coauthor Fabio Parasecoli for stimulating discussions on food, culture, and design, for our shared and exciting ethnographic adventures in the field, for all I have discovered teaching and prototyping together with students in classrooms and design studios, and for his unwavering support for my academic work, especially when the going got tough. Thank you, Fabio, for your friendship and openness to our working across disciplines and experiences.

I want to thank my family and friends, whose love, companionship, and care has sustained me as I worked on this and other projects these last two years in Warsaw: Edgar Bąk, Paulina Wróbel, Mirek Filiciak, Hanka Samson, Luka Rayski, Łukasz Zaremba, Magda Szcześniak, Ola Rayska, Marcin Wicha, and Ewa and Andrzej Halawa. Thinking about Global Brooklyn comes from personal experiences as well: Mirek brews a perfect drip from Ethiopian beans when we work together in Poland, Luka is always up for bar hopping whether in Greenpoint, Bushwick, or Warsaw's Śródmieście. I spent so many hours talking to Kasia Samson at Warsaw's Relaks, a café which makes an appearance in the intro (they made my bike too!), and hanging out with Randi Irwin at and around Brooklyn's 61 Local (see Chapter 8), where on my last visit we toasted to her marriage to Holly Creenaune. Thinking, writing, editing, and teaching is never a solo act, and I like my work to be sensual and convivial. One philosopher considered only those ideas and debates worthy that could "dance," and in this spirit I want to also acknowledge both casual and festive cooking, eating, drinking, and talking with, in no particular order, Leilah Vevaina, Tim Rosenkranz, Scott Brown, Kylie Benton-Connell, Soo-Young Kim, Agnieszka Jacobson-Cielecka, Maciej Siuda, Łukasz Bluszcz, Zuza Rudzińska-Bluszcz, Kaja Sztandar-Sztanderska, Maria Reimann, Marianna Grzywaczewska, Karol Murlak, Kasia Boni, Marta Olcoń-Kubicka, Marcin Serafin, Zosia Boni, Małgorzata Bakalarz-Duverger, Piotr Łukasiewicz, Paweł Ciacek, Adriana Prodeus, Roch Dunin-Wąsowicz, Mikołaj Lewicki, Feliks Tuszko, Filip Katner, and Greg Podleśny.

This volume draws on my work at the intersection of humanities, social sciences, and design, which began at the Parsons School of Design and benefited greatly from the friendship and inspiration of Susan Yelavich and Shana Agid. Design then took me to New School's GIDEST, where I was lucky to start working with Fabio and continued to learn from Hugh Raffles, together with an exceptional cohort of fellow graduate students and more advanced scholars. Most of what I know about design, however, I have learned from my friends, colleagues, and students at the School of Form design program, including Jola Starzak, Dawid Strębicki, Arek Szwed, Bartek Grześkowiak, Honza Zamojski, Monika Wietrzyńska, Zygmunt Borawski, Monika Rosińska, Jarek Hulbój, Paulina Matusiak, Mateusz Falkowski, Dorota Kabała, Dawid Wiener, Li Edelkoort, Ewa Klekot, Filip Zagórski, Monika Jakubiak, Wojciech Dziedzic, Maldoror, Olga Milczyńska, Anita Basińska, Oskar Zięta, Jagna Jaworowska, and Beata Wilczek.

—*Mateusz Halawa*

Part I

Finding Global Brooklyn

Introduction

Global Brooklyn: How Instagram and Postindustrial Design Are Shaping How We Eat

Mateusz Halawa and Fabio Parasecoli

It is an August morning in Warsaw, finally balmy and warm after a few days of rain and grey skies. We are waiting for our coffee order in Relaks, a café in the long-prestigious neighborhood of Stary Mokotów, now going through a new wave of postsocialist gentrification. To the right of the entrance young, tattooed, and studded baristas work in front of the blackboard-painted wall listing in deliberate but quirky white chalk lettering the requisite kinds of coffee, including the drip, the chemex, the flat white, the cascara, as well as particular sources and regions of the beans. In the main space on the left, people socialize, write on their laptops, and take work meetings.

Designed by local architects for an owner who is a designer himself, this coffee place makes good use of the old 1970s wood paneling, the newly chic 1960s socialist designer wood tables, and some cheap IKEA couches. Relaks is recognizably cosmopolitan, with all the paraphernalia of the global coffee cult—the Hario V60 drip and filters, La Marzocco coffee machine, and an extensive collection of pieces from the Polish poster school. The café is also deeply local, a significant point for the identity of the neighborhood as it is experienced by its younger inhabitants, often children of the local intelligentsia. Relaks is both a center of community life and a node connecting Warsaw to Berlin, with its much appreciated roastery, The Barn, and to Brooklyn, which arguably originates many of the urban trends that can be observed here. A MacBook third space outside of the home and the office, Relaks is carefully designed to be distinctive—and while the place and its baristas sometimes make an appearance in events around the city, it is not a concept that would lend itself to corporate-style branding, scaling up, or turning into a franchise. The aesthetics speak of uniqueness, even idiosyncrasy, as it happens in similar places that increasingly populate the world.

The roasting and brewing, which are a spectacle in itself, embrace the organoleptic characteristics of third-wave coffee, with its focus on provenance, its preference for acidity and light roasts, and its interest in respecting and even highlighting the fruity and floral notes of the beans. As baristas participate in tastings, or "cuppings," they generate complex vocabularies describing taste profiles that are appreciated by the connoisseurs but not the popular palate. "I'll have the Ethiopian," goes a typical order. The flavors themselves may be puzzling to many who grew up appreciating Arabica-heavy dark roasts. Brews described as "bright" may taste downright funky and, in the words of one San Francisco-based artisan roaster, be admittedly "tough to drink," but nevertheless "really interesting" (Deseran 2013). These young coffee enthusiasts talk taste with the sophistication and a sense of belonging, which the Western bourgeoisie have historically reserved for wine. There also exists an unarticulated list of no-nos, which makes an appearance when someone uninitiated shows up: no espresso in a paper to-go cup; no Americano, but drip instead; no syrups or whipped creams. It is the anti-Starbucks: choices are limited and well curated, while the flavor and the preparation are more barista-centric than consumer-centric. This is a service economy with a Puritan streak, communicating what should *not* be done here.

Before renovation some years ago, the café was connected to a bike shop and the fixed-gear crowd still often makes an appearance. Relaks celebrates manual labor and allows the guests, largely employed in postindustrial, creative, and service economy sectors, to fantasize about the life of manufacture. The figure of the barista is exemplary here: her decidedly manual labor does not remain unseen and unexamined like it would some years ago. On the contrary, through training, self-teaching, competitions, and storytelling, it becomes publicly visible as a valuable and fashionable practice. Her craft is also celebrated and sought after as a form of knowledge and expertise, a spectacle to be watched and worth the wait.

Although not performed directly in front of customers, expert craft is also central to the success of Open Baladin in Rome, a restaurant that was a pioneer when it opened with its focus on artisanal beers, in particular Italian ones. Baladin is among the first craft beer brands in Italy, an offshoot of a pub in Cuneo, Piedmont, that turned in 1996 into a "brewpub" where young brewmasters started experimenting and trying new flavors and techniques. Embracing their motto "taste in evolution," the brewery has acquired national and international renown, providing a business model for many other craft beer producers in Italy. They consider their production plant a place for research and creativity, embracing collaboration as an opportunity for learning, teaching, and the cross-pollination of ideas, styles, and techniques. The same inspiration informs their restaurant in Rome. Its website tells us that the place "in all its aspects, starting from our beer selection and food menu, the care with which we educate our staff and the look of our bars—is a statement, an 'open letter' to all the people who still love good and genuine products coming from the earth and elaborated by human enthusiasm and creativity" (Open Baladin 2019).

Although the environment seems laid back and unpretentious, every detail is clearly carefully designed. The largest space is dominated by a wall with wooden shelves where bottles of craft beers are displayed. Right in front of it, the long counter with many beer taps showcases the variety of what is available. The other walls are decorated by colorful murals and a long blackboard-like black strip on which the names and descriptions of the available beers are written in brightly colored chalk. The furniture is simple, made of light-colored wood, metal, and leather, vaguely alluding to an industrial workspace. Although well-coordinated and custom-designed, the refusal of unnecessary embellishments conveys a sense of straightforwardness. On each table, a square piece of metal holds a small tin bucket that contains paper towels. The lighting fixtures are also minimalistic: naked Edison bulbs over the counter, simple covers on the tables.

The menu presents a mix of Italian dishes (pasta and risotto) and foreign specialties such as burgers and falafel. The common element is the attention to the quality and provenance of the ingredients. The element of craft in the kitchen is central to the identity of the restaurant, as well as its search for authenticity in ingredients and techniques.

The centrality of such themes and concerns is at times not even tied to specific places but lives in the digital space of social media. Browsing our Instagram feed, we may spot an image of a nice café counter with a whimsically written menu, behind which a bearded youth in white shirt, suspenders, and bow tie carefully pours hot water on coffee grinds, against a background of potted monstera and other plants hanging from pipes and prominent structural elements. We can almost smell the aroma of the coffee, unconsciously getting ready to hold the warm cup in our hands and start sipping. All the elements in the picture redundantly convey the same feeling of care, labor-intensive attention to detail, coziness, and comfort, reflecting an approach to food service that does away with formalities and focuses on what counts: the experience of the customers (and, in this case, of the viewer as well). Instagram posts of morning coffee aestheticize the everyday, elevating what was earlier mundane to the status of a sensuous event and at times even art. Except that we may be hard-pressed to figure out where the picture was actually taken, as similar surroundings and scenes could be enjoyed in far-flung locations.

Although it exudes a sense of place and presence, in reality the shot could have been taken anywhere in the world, a sensation that is intensified by the capacity of hashtags like #specialtycoffee to arrange next to each other identically marked and similarly staged images from around the world. Whoever took the photograph may have felt the attraction of this deterritorialization as well, as they inserted themselves into a cosmopolitan flow of styles and sensibilities. We may get more specific information from the geolocation conveniently provided by the application: a name would help us put what we see in context, perhaps even "follow" the place and visit it while we travel. Accidentally, the same information can be scrubbed by algorithms that provide advertisers with useful information about how trends are taking place, who is looking at them, and where. The barista in the picture, the photographer who posted the picture, and those who look at it are all turned into data points with great economic value (Zuboff 2019). We may not really care about all this, though, as we scroll to the next picture, the next scene, the next amazing coffee, wonderfully brewed in cups that are explicitly designed to enhance the complexity and richness of its scent.

Life in Global Brooklyn

The vignettes we provided are quite likely to generate a sense of déjà vu among readers who may have had comparable and at times strikingly similar experiences around food and drink in Buenos Aires, Amsterdam, or Kampala. Regardless of the context, people congregate around and talk about food and drink in ways that have become recognizable as global trends. In this book, we explore the contours of this transnational aesthetic and at times ethical regime, which we have called "Global Brooklyn." It is a cultural formation constituted by a recurring, loosely codified set of material objects, constructed environments, practices, and discourses that may or may not appear at the same time, in similar patterns, or even with the same meanings (Parasecoli 2016; Halawa and Parasecoli 2019). We have observed it in cafés, restaurants, and stores, and identified it in reports from collaborators in cities worldwide, including New York City, Warsaw, Rio de Janeiro, Chiangmai, and Mumbai, among others.

After examining the emergence of the phenomenon, this introduction will describe Global Brooklyn's constitutive elements in terms of an ideal type, an approach previously used effectively in assessing food consumption patterns in urban spaces (Irvin 2016). We then compare Global Brooklyn's worldwide circulation with similar food-related transnational cultural formations—from sushi to Starbucks—to identify what is distinctive. We subsequently discuss the actors involved, suggesting hypotheses about the reasons for the success of Global Brooklyn around the world. Finally, we provide a narrative of how the project came to be and how it developed into a multisited collaboration among scholars and practitioners from very diverse backgrounds, whose contributions will be briefly introduced.

In Madrid, Montreal, or São Paulo, the aesthetics and the sensory landscapes connected with what we call Global Brooklyn are frequently mentioned with reference to the New York borough, which for better or worse has raised to worldwide fame as one of the main epicenters of and models for food-related trends, whether they actually originated there or not. Part of Brooklyn's visibility is its symbolic role as the anti-Manhattan, a place where immigrants and working-class groups of various ethnicities were able to make a living (LeBesco and Naccarato 2015). If the public perception of Manhattan speaks to an outdated "world-is-flat" imaginary of globalization, contemporary Brooklyn suggests an alter-global, diverse mode of worldliness. Moreover, in the late 1990s and the early 2000s Brooklyn offered comparatively affordable rents to both entrepreneurs and their customers, facilitating the emergence of enclaves where Global Brooklyn-style establishments thrived, often located in abandoned industrial buildings and less than desirable neighborhoods.

Inevitably, Brooklyn's reputation is built on the global pervasiveness of US media, from TV to cinema, that made the location a recognizable point of reference for popular culture worldwide. It is an imagined territory rather than an actual place; the less savory realities of gentrification, ethnic tensions, unemployment, and urban decay are strategically written out to allow consumers around the world to superimpose their desires, aspirations, and preferences. These are themselves inevitably the result of negotiations between local sociocultural contexts and the circulating imaginaries provided by US media companies. The New York City borough has become central in the contemporary collective imagination, especially for generations that aspire to embody the same coolness and trendiness they see projected onto it.

The Global Brooklyn we evoke in this book is a floating signifier, whose always-changing meanings and relevance within a cultural system are never fixed but rather constantly negotiated among old and emerging centers of interest and forces vying for cultural hegemony (Laclau and Mouffe 1985: 93–148). We are fully aware that the origins of what we discuss are not directly connected with the New York City borough but are much more complex and dispersed. Among the cities that contributed to its elaboration, we can mention Portland, Oregon, with its crafty, ethical project of good living that is so recognizable—at least in the United States—to become the target of a whole comedy TV show, *Portlandia*, which acutely but affectionately takes jabs at alternative and hipster cultures. In San Francisco, digital and tech culture, a central source of income, generated appreciation for postindustrial materiality, maker culture, and fix-your-bike DIY attitudes. The availability of money among millennials employed in the tech sector supported cafés and other establishments where nomadic clients could sit with their laptops and work. Seattle contributed with its role in the growth of coffee culture in the United States, and it offered some interesting experiments in creating alternative models of food consumption, particularly in the punk scene (Clark 2004).

Berlin, once described by its own mayor as "poor but sexy," was also crucial in generating the aesthetics of repurposed, beautiful trash, which embraced the peripheral and reflected its complicated history of cosmopolitanism. As a matter of fact, in Central and Eastern Europe, from Zurich to Warsaw, Berlin may be a more common and understandable reference than Brooklyn, although New York City is still relevant in the shared imaginary. Furthermore, the connection between the aesthetics of Global Brooklyn and the visual approach of the New Nordic Cuisine appears closely related, raising questions about what came first, for instance, in Copenhagen. All these cities probably played a role as important as the New York City borough in shaping approaches to consumption that constituted a critique of mass production and contributed to the development of the aesthetics and practices of Global Brooklyn, connected with dynamics of gentrification and the North American "food movement" (Finn 2017). Following these reflections, we will use "Global Brooklyn" not to refer to the specificities of a precise place or to claim a single origin for what we are exploring, but rather to describe an aesthetic and a way of consuming food and drink that popular culture and mainstream media have frequently identified with the New York location.

While academic researchers may have overlooked it, lifestyle media have picked up on this new visual landscape, from fonts to clothes to interiors, but offered little analysis, all while reinforcing their identification with Brooklyn. They mention "industrial furniture, stripped floors and Edison bulbs" (Chayka 2016), reclaimed wood, and refurbished industrial lighting. Kyle Chayka (2016) has written of "AirSpaces" and connected them to wealthy mobile or telecommuting elites and their precarious workers. Writers rightly suggest a knowing, or knowledge-intensive, aspect to these new spaces, which are very self-aware, responsive to the unending stream of their representations and reflections, and inviting more. The current Brooklyn aesthetic, wrote the New York Times, "if it exists, has a certain intellectualism to it," and frames the everyday in terms of a social reality that is more "about day-to-day coping and nesting than peacocklike display" (Chang 2012). The website QZ quipped:

> The "Brooklyn of (insert any city)" has become the global nomad's shorthand for any town's trendiest, most visually eclectic, and often most clearly gentrifying areas. Such enclaves include the suburb of Pantin of Paris, Shimokitazawa neighborhood in Tokyo, Florentin in Tel Aviv and Shoreditch in East London. What the Brooklyns of the world have in common is an aesthetic: repurposed décor from mixed sources and a palpable anti-slick, anti-corporate sensibility in favor of nostalgia for the 19th and early 20th centuries. (Quito 2016)

Focusing instead on the United States, the website Thrillist proposed a list of the Brooklyns in every state: "We rounded up the most Brooklyn 'hood in every single state, based on metrics including trendy restaurants, 'craft' cocktail bars, bike friendliness, and, of course, urban expansion. By the time you finish reading the list, they probably won't be cool anymore, so hurry up." There is always a time lag between the most current trend and what was trendy just a short time ago. While Global Brooklyn may feel a little passé in some places, in others it is perceived as a relatively new phenomenon, eliciting conversations that closely remind of those from the locations where the aesthetic regime has been present for a longer time. In 2019 the newspaper *El País* in Spain wondered: "Why do restaurants look increasingly similar? Vintage look, posters and blackboard with supposedly handwritten letters, shared wood tables, desk chairs . . . maybe restaurants are becoming all the same?" (Couceiro 2019).

As these spaces proliferate across global cities, they tend to be discussed in the language of authenticity, craft, gentrification, and the ever-shifting and hard-to-define "hipsterism," of which Brooklyn is at times (reductively) considered an expression. Elsewhere in the world these restaurants and cafés instead embody cultural and social aspirations, expressing entrepreneurial and consumer desires to become part of the cosmopolitan middle classes and to be integrated with global trends, coded as both exciting and prestigious. As the chapters and the dispatches in this volume indicate, apparently similar sensoria may be activated in vastly different ways, reflecting heterogeneous priorities and values.

Toward the Ideal Type of Global Brooklyn

While manifestations are context-dependent and may articulate only some elements of what we describe, there exists, we argue, an underlying and coherent regime that governs the more or less spontaneous appearances of Global Brooklyn. This may be understood through the Weberian method of the ideal type. An ideal type identifies and selects recurring features and elements across observations that may or may not be present in all instances. Although abstract and hypothetical, by systematizing and ordering the chaos of particularities, an ideal type constitutes an effective analytical tool to better interpret the underlying dynamics and determinant factors of the overall phenomena and to assess the particular cases one may encounter. We draw on Weber's (1997) concept for its capacity of generating insights that are both interpretive of specific situations and comparative between dispersed locales. A strategic fiction useful for gaining novel outlooks on the real, a hypothesis worth considering, Global Brooklyn for the purposes of this volume is an ideal-typical construct, because all of its characteristics will not always be equally present in each particular case. As we have worked with our collaborators around the world, it was also an ideal type *in progress*, offered up for critique and refinement.

The various manifestations of the Global Brooklyn ideal type tend to cohere around interconnected yet distinguishable axes, which we explore in this section. These develop through discursive and interactive practices around objects and spaces that blur the distinction between the symbolic and the material. They provide the flexible and shifting principles of a generative grammar, a loose combinatory that allows for the emergence of local manifestations building on the same shared repository of elements that can be assembled in very different combinations. This grammar is coherent enough to allow manifestations that are recognizable in different contexts and by different agents, even if they are understood and framed in notably different ways. This points to a constant shift of meaning in Global Brooklyn, ranging from a profound and felt engagement with its more ethical elements to the calculated exploitation of a trendy design approach meant to reap commercial benefits. Hotel chains, as well as singular boutique hotels, have become among the most enthusiastic adopters of the aesthetic regime (perhaps under pressure from "authenticity" marketed by Airbnb), often decoupled from any other elements identified in the ideal type.

The collaborative work collected in this volume began by looking at the ideal type from the perspective of the following five axes.

Designed Experiences

First, there is design. Eating and drinking in Global Brooklyn are thoroughly designed experiences, which often seek to hide their artifice behind a performance of authenticity. In this context, we use authenticity to denote the expected adherence of ingredients, dishes, and practices to an idealized model based on integrity and genuineness. Authenticity is a cultural and ideological effect of ingesting products framed as unadulterated, honest-to-goodness, and with connections to real people and their stories (Parasecoli 2017). Of course, authenticity is socially constructed and highly contextual (Zukin 2009). It may refer to personal or communal experiences; expectations based on perceptions, external information, or even bias; and in certain cases the connection with cultural or ethnic groups that may be considered as less affected by modernity and globalization, possibly stuck in an imagined unchanging present that also coincide with their past.

In the realm of objects and sensory environments, the desire for authenticity reflects a critique of artifice as a reflection of corporate, repetitive, lowest-common-denominator, consumerist culture. Even in the "ready-made" reclaimed spaces where we see little architectural intervention, Global Brooklyn relies on an intense work of "designing the invisible" such as services and interactions (Penin 2018), as well as on graphic and communication design. This includes creating digitized representations and scripting interactions as the experience is enhanced by partly moving it online, as we discuss later. Whether professional or not, the designer is a key intermediary in the making of Global Brooklyn. Its look and feel is heavily postindustrial, reflecting a preference for repurposed material, urban flotsam that is revaluated and elevated through reassemblage. Reclaimed wood and euro pallets used as building blocks manifest a growing interest in circular economies and materialize a critique of unsustainable development, while rusty iron embodies a nostalgia for a less virtual world. Old vehicles are refurbished into rugged but cool food trucks. Stripped down, often retrofitted architecture goes together with old objects taken out of their original contexts, including mason jars as containers for drinks. There is a preference for the low tech and a deliberate rejection of most current innovations: Edison bulbs hung over vinyl record players (Bartmanski and Woodward 2018). There are constant allusions to unique objects salvaged from the mass-production past that highlight the uniqueness and originality of the food offering, at times with a tinge of nostalgia (Jordan 2015). What used to be necessity, is not a choice, generating a tension between the logic undertones of sustainability and emerging forms of bourgeois antiquing applied to a new class of things that earlier were not seen as possessing "heritage" value.

The visibility of infrastructure like pipes and ducts provide a form of urban authenticity stripped to the core, suggesting that this new scenography is not scenography at all. Even though Global Brooklyn establishments have a presence in social media and routinely offer free Wi-Fi, a preference for the analog over the digital—a celebration of glitches, imperfections, material obduracy, and slowness—expresses a longing for uniqueness and serendipity largely eliminated by platform capitalism. Through visual references and carefully staged performances, manual labor is elevated into craftsmanship whose value is enriched by new meaning, higher social status, and cultural capital (Ocejo 2017). Menus are handwritten on blackboards to allow daily changes (or suggest they are made), with quirky "hipster sans serif" lettering and illustrations (Usborne 2014). To the suspicion of critics: the hipster moment "did not yield a great literature," but did make "good use of fonts" (Greif 2016: 220). A significant aspect of these designerly practices is the interaction between people and materialities, be their wooden planks, coffee grounds, or kombucha's cultures of bacteria and yeast. Global Brooklyn intermediaries are in this sense not workers or entrepreneurs, but *makers*, who understand and often romanticize what they do in terms of world-making: an ethical, life-transforming, constructive, and embodied practice. This new language of making resists the easy separations between the head and the hand in framing labor (Ingold 2013; Sennett 2009).

A Sensory Regime

Global Brooklyn is a sensory regime that expresses a renewed relationship with locality and provenance as a reaction against the anonymity of industrial production and mass-marketed food. Organoleptic traits, from acidity in coffee to oxidation and strong smells in natural wine, which in the past would have been considered defects, are now expressions of a more direct relationship with raw materials and places of origin. Kombucha fermentation connects drinking to alternative health practices, make funky tastes appealing, and create a lively sociality as people mix offline and online interactions to share their starters, or "scobys" for the initiated, trade stories of domestic experimentation, or recommend other artisans. Natural wines convey yeast scents and other aromas that are interpreted as markers of the places where grapes are grown, the must is fermented, and the wine is bottled and aged (filtering can compromise all this, and, as such, should be avoided). Craft beers may see the addition of fruits, berries, and other ingredients that connect the consumer with the place of production, the creativity of the brewer master, and the business acumen of the brewery. Coffee has to retain and highlight the aromatic notes of the beans from which it is brewed, showcasing the roasters' skills in the selection and handling of single-origin varieties. Textures and shapes of things are rugged and reveal the process of their making, whether it is "pulled pork" with mashed avocado, visible weld lines in metal furniture, or a needle skipping over a dusty vinyl record. Such resonances across different material domains lend themselves to the analysis of Global Brooklyn in terms of "qualia" or "qualisigns of value," in the sense that the physical qualities of objects and environments not only express discursive meaning but are also charged affectively, influencing the emotional aspects of experiences (Chumley 2013; Chumley and Harkness 2013; Harkness 2015).

Food and their surroundings become arenas in which identities can be constructed, negotiated, and expressed through conspicuous—although relatively affordable—consumption. The new sensory preferences can also be interpreted as an expression of desire for distinction, in terms not only of cultural capital but also of generational tensions, so that what older consumers would perceive as the height of quality can now be construed as old and démodé. While changes in aesthetic categories of degustation have a direct impact on embodied sensory practices, they also bring about economic consequences, as customers are willing to pay a premium to have access to those experiences. As we will see, a certain level of knowledge and cultural capital is necessary for customers to understand and enjoy the experience, justifying higher costs. Formerly mundane dishes are elevated in price and status, if not always in quality, undergoing processes of valuation that generate entire new industries. Valuation is far from being an exclusively cognitive process of interpreting and giving meaning to what is already there. It is also an active and intentional process that takes place while making and imagining what is not there yet (Heuts and Mol 2013; Lamont 2012). This last aspect is another reason why the sensory dimension is so strongly connected to design practice with its drive to innovation and methodologies of prototyping.

Networked Communication

The third significant axis for our analysis is communication. Global Brooklyn is the foodway of the network society (Dürrschmidt and Kautt 2019; Rousseau 2012; De Solier 2013; Lewis 2020). Social media facilitate the transmission and adoption of this cultural formation, unfolding in digitally augmented spaces within deterritorialized communities of practice, which share and learn from each other online and create new criteria of value. In Poland, craft beer enthusiasts post clips on YouTube and exchange recipes and tips in the comments. Your pulled pork sandwich in Los Angeles may well be assembled from local pork, but it is equally a product of Weibo or Facebook, from butchering techniques and cuts to the use of spices. The circulation of Global Brooklyn takes place mostly through visual means, in particular through pictures and videos. Spaces and meals are designed in response to the global circulation of images, values, and ideas deployed into local things, and then are in turn further photographed, discussed, and disseminated on Instagram and pinned on Pinterest. They become part of a coherent aesthetic that may nevertheless sometimes look out of place in their physical context. However, we cannot forget that there is a profound offline effect to this communication: a good cup in a Warsaw Global Brooklyn café may require gathering coffee in Ethiopia, then roasting it in Denmark, and brewing it in Poland with coffee filters shipped from Japan. Online communication mobilizes certain fetish commodities and objects that travel increasingly long distances, sometimes putting Global Brooklyn at an awkward position vis-à-vis its claims to sustainability. On some level, the focus on valuable imports carrying strong cultural and cosmopolitan associations renders some of Global Brooklyn outposts a twenty-first-century version of what twentieth-century Europe knew as "colonial stores," with their spices, cocoa, and coffee coming for afar. It is not surprising, then, that so many contemporary cool bars or riverside pizzerias are constructed from the repurposed detritus of the infrastructure of the global commodity chain, and not only the long-known commodity burlap coffee sacks but also highly modern wooden euro pallets that fit the TIR trucking standards and entire cargo containers.

Contemporary social networks and digital technologies are used not only for marketing and promotion but also to establish cultural claims and mediate social interactions. Consumers are asked to co-produce value by photographing food, geolocating themselves, tagging friends, and sharing images. The material they generate is immediately nostalgic: the moment you live it, you mediate it; you gram it and write about it, and it becomes old (Halawa 2011; Schwarz 2009). As participants in Global Brooklyn document everything all the time, what they record is always already there to be archived, creating a sense of nostalgia for the present (Jameson 1991, see Appadurai 1996: 29–30).

It is not by chance that at the same time analog instant cameras are back en vogue, just like typewriters and other symbols of older forms of communication. This makes the analog more desirable than the fleeting digital media, which are also reflected, supported, amplified and, in a certain sense, legitimized by coffee table volumes, cookbooks, and magazines. Paradoxically, the interest of Global Brooklyn actors in the analog has extended to printed materials, which had been prematurely declared dead in the wake of digital food writing and reporting online. Well-designed and creatively edited magazines printed on high-grammage matte paper such as *Usta* (Mouth), *Zwykłe Życie* (Ordinary Life), and *Kraft* (Craft) in Poland, *Cook Inc.* in Italy, and the now defunct *Lucky Peach* in the United States, have all contributed to a revival of the genre in an upscale and refined version. This constant production of shared information is enmeshed in media outlets' frantic quest for consumers' attention, which is increasingly treated as a scarce resource and as a commodity. As a consequence, expensive juice cleanses find their match in a new puritanical discourse of "digital detox." Authenticity is sought in the promise of a life-changing disconnection, a Waldenseque exit from the online to the offline, as if an offline dimension totally unplugged from the network could actually exist, in the time of the Internet of Things (Syvertsen and Enli 2019).

Expertise and Skills

The fourth dimension of the Global Brooklyn ideal type is the role of reflexively deployed expertise and skills. This cultural formation hinges on knowledge-intensive practices of the highly educated who strategically mobilize marketing, design, and social critique to create and control it. Working is increasingly networking; relationships and attachments are more and more expressly an object of cognitive, relational, and emotional labor. From this perspective, Global Brooklyn is co-created by tightly connected publics of consumers, influencers, and providers, equal parts reclaimed wood and digital platforms. For all its appeals to spontaneity, roughness, and authenticity, it is thoroughly intellectualized and self-aware, like the Brooklyn graduate school lumberjacks. Its cultural status is a problem not only for those who observe and describe it from the outside but also for the insiders who endlessly debate, tweak, experiment on, and lampoon its various aspects. Interactions include storytelling, tasting, and what Arjun Appadurai (1986: 22) calls "tournaments of value," which are highly ritualized competitive events like specialty coffee cuppings, including championships organized by country, region, up to "world cup tasters." Other interactions, online and off, include professional and advanced amateur training, networks-building with other crafts persons, and education of the customers. Entrepreneurs, chefs, and consumers come to share a common language, expectations, and values that reveal new forms of cosmopolitanism hinging not only on the consumption of foods from elsewhere but also on pride and passion for local, traditional specialties (Parasecoli and Halawa 2018). Customers often become so passionate that they try their hand at production in forms of amateurism that are constantly aspiring to professional recognition and paradoxically tends to institute hierarchies of taste and knowledge through participation in workshops and training classes, as well as acquisition of information through books and other media. Amateurs quickly feel entitled to look down on others at the coffee place or at the natural wine shop. In this sense Global Brooklyn reveals an anti-democratic streak that would seem to counter the supposed desire for equality often ascribed to foodies (Johnston and Baumann 2015).

In Global Brooklyn, your sandwich guy may think of himself as a personal brand, trying to "engage" and "influence" you on Instagram through his expertise and stories. This approach was well satirized in a *Portlandia* sketch in which a waitress narrated the life story of a free-range chicken, providing detailed information about it and even its picture. In cafés, ad hoc cuppings often become an opportunity for strangers to discuss flavor and create a temporary affective community, a necessary component within the shift in aesthetic categories we discussed above. These are coffee places with a *theory* of good coffee and food trucks with an *ethos* attempting to change the world—both writ large, as when engaging with global issues of sustainability, and writ small, as framed through notions of community or neighborhood. Such strategies are aimed at achieving success in the network economy, very often measured in relatively novel terms of social, environmental, or justice impact. However, they largely keep themselves at a distance from direct engagement in social movements, revealing a constant tension between references to grand narratives and the care reserved for small radius issues. This economy, also termed virtual or reflexive, is in Lisa Adkins's (2005) terms knowledge-intensive and service-intensive, as a great capacity of added value comes from practices of designing, branding, and emoting. As the economy becomes more virtual and the production of value more abstract and removed from what we have associated with real economy (think the rise of speculative finance), Adkins (2005: 111) notes a paradox in which "a greater emphasis is being placed on issues of embodied performance and the significance of human or physical capital appears to be intensifying."

A Revaluation of Artisanal Labor

Many in Global Brooklyn appeal to the ethos of manual labor, which constitutes the fifth and last dimension of the proposed ideal type. It celebrates authenticity and craft, previously displaced and rendered anachronistic by both the fact of industrial production and the cultural excitement surrounding modern food engineering. Both economically and culturally, what is often at stake in this new moment of Global Brooklyn is the desire of middle-class knowledge-workers to perform typically lower class forms of manual labor such as butchering, cooking, and brewing. As the sensibility of popular culture has moved from the glitzy Manhattan of *Sex and the City* to the grimy Greenpoint of *Girls* and beyond, Global Brooklyn is increasingly populated by individuals who have dropped the dreams of glamour for a more gritty aesthetic. The DIY practices and experiences are expressions of agency and autonomy: makers want to find their own way, without necessarily following the rules set before them by old-timers. However, they accept direct transmission of knowledge through exchanges with peers or immediate predecessors whose experience they recognize and appreciate. This seems to resonate with a shift in the regimes of mediation: these are not hierarchical-pedagogic book learners, but a networked swarm of YouTube peers and tastemakers.

In Global Brooklyn, the staff, wearing simple or no uniforms, show relaxed manners and a certain lack of hurriedness, partly to underline the artisanal, bespoke, personalized style that customers of those establishments are supposed to appreciate. Service, often informal and at times idiosyncratic, is nevertheless carefully thought out and curated. It results from business decisions that may reflect ethical choices, approaches to labor, and connections with larger concerns with the food system such as sustainability, fairness, and health. These approaches sometimes provide a critique of contemporary consumption regimes by articulating desires for social justice and alternative arrangements. The consequence of the engagement with social and political matters is that, despite the easygoing atmosphere, some business owners and staff may take their jobs quite seriously, often framing them in terms of mission, as the manifestos increasingly appearing on walls, menus, and websites seem to indicate. While scholars and activists have been operating at the intersection of food and politics for a long time, Global Brooklyn mainstreams these practices while risking their cooptation by the same logics they had initially sought to transform. In fact, in some of its manifestations, Global Brooklyn does appear quite devoid of any social or ethical concern, being merely used as a trendy aesthetic to attract customers.

Global Brooklyn and Other Modes of Diffusion

By creating globally repeating forms of urban locality, Global Brooklyn positions itself in a long history of food-related cultural formations that have shown the same scope of global dispersion, such as sushi, French fine dining, supermarkets, and various transnational corporate chains à la McDonald's. Is Global Brooklyn just another expression of established dynamics of culinary soft colonization, or does it present any qualitatively different elements?

Just like Global Brooklyn, sushi restaurants, which have been successful in Western countries from the late 1980s, present a specific and highly recognizable sensorium: a spare interior design style with a prevalence of wood and other natural materials such as paper and stone; distinctive Japanese tableware and chopsticks; condiments that require some explanation, both in terms of flavor and usage. There is a sushi master performing highly specialized skills in front of patrons, who in turn are willing to engage with new aesthetic judgments, discourses, and practices about what can be considered edible, palatable, and fresh. Unlike in Global Brooklyn, sushi restaurants in the West relied on an underlying global supply chain, which for high-end establishments famously hinged on a singular point on the globe, the Tsukiji market in Tokyo (Bestor 2004; Issenberg 2014). The restaurants were also predicated on the presence of specialized labor that was supposed to be trained in Japan, upholding the tradition of master-to-apprentice transmission of knowledge and skills. As it happens in many kinds of Global Brooklyn establishments, acquisition of knowledge was necessary for patrons to fully enjoy the experience, which in turn became a token of cultural capital.

The dynamics of the global dispersion of sushi share elements with the success of French cuisine in fine dining from the end of the nineteenth century, as systematized by Carême and Escoffier. The exportable cultural formation included a training method built on discrete and successive learning blocks such as knife skills, stocks, and sauce making.[1] Such a method required chefs to train in the French apprenticeship system, which over time created a tight network of traveling labor. French fine dining also included a specific model of kitchen organization, the brigade, and a set of ingredients that had to be imported from France, such as champagne, foie gras, truffles, and traditional cheeses. Like sushi, French fine dining required a certain amount of cultural competence from customers in order to understand menus, order dishes and wine, and even be able to properly use tableware and glasses.

From the point of view of circulation of labor and reproduction of skills and craft, Global Brooklyn is much more flexible and porous than sushi or French fine dining. As there is no singular geographical center of diffusion, knowledge is produced and transmitted from different locations and through diverse methods. This flexibility also lessens the need to respect canons and the "wisdom of the elders," although as we observed earlier this does not stop some amateurs from attempting to create new canons and hierarchies with which to discipline those not in the know. As a result, actors in Global Brooklyn may want to learn from others that they personally appreciate and respect; however, they feel free to experiment and to learn skills from peers and even learn and share as they go in online communities of practice.

The global dispersion of cultural formations we have discussed so far may have real (Japan, France) or imagined (Brooklyn) centers of diffusion, but lacks any forms of centralized coordination from an economic and communication point of view. This constitutes a substantial difference with the spread of supermarkets from the United States, which presented a mixed mode between centralized top-down diffusion and autonomy of networked establishments often owned by individuals. Supermarkets also offered a recognizable but adaptable sensorium, which included ways for customers to use space over time, new self-service designs, and a novel presentation of products using merchandizing approaches (Cochoy 2010).

The corporate model of commercial expansion of Pizza Hut and Starbucks appears even less flexible, embracing a corporate approach of top-down diffusion, with decisions made in boardrooms and designs produced in agencies, mostly located in the Global North. The logistics, the supply networks, the business model, and the sensorium remain closely monitored and coordinated by a central authority. The corporate structure enforces very strict aesthetic regimes, in which even the smallest element of the physical environment is predetermined, together with service and space use modalities. This is also the case for glitzy flagships stores, such as the spectacular Starbucks Reserve Roastery in Milan, where uniqueness is designed down to the smallest element to elevate the company brand and ultimately serve the entire coherent aesthetic. All decisions about material elements, practices, and communications are centrally made to ensure that customers have a uniform experience wherever they are around the globe. Granted, some flexibility is provided in terms of menus, creating the dynamics sometimes referred to as glocalization: McDonald's in Morocco may offer a McArabia (pita with lamb meat) and those in Poland a WieśMac (a burger with mustard and horseradish sauce).

Comparing the dispersal of Global Brooklyn to previous examples, we observe a shift from linear, corporate, top-down broadcasting models of the transmission to network dynamics. Each location turns into a communication node and everybody receives, elaborates, and distributes elements without any apparent decisional center or mysterious nonhuman authority of black boxes in the form of protocols and algorithms (Zuboff 2019; Beer 2019; Bucher 2018). As each node is potentially connected with all the other nodes, with only a loose sense of cultural hierarchy, communication becomes multidirectional and diffused, with signifying elements left open to diverse interpretations. However, it is important not to romanticize these aspects and to reflect about these supposed networks in terms of political economies. Network dynamics inevitably create nodes with more power, where privileged cultural and economic intermediaries operate, and establish core-periphery relationships. As much as the rhetoric of Global Brooklyn frequently insists on peer-to-peer connections, inequalities cannot be totally avoided. At the same time, the decentralized, distributed, image-centric mode of circulation and mediation of Global Brooklyn promotes forms of marketing that, although still aimed at increasing visibility and sales, also place a greater value on consumers, who by being elevated to the status of "influencers" or "personal brands" become themselves nodes in co-creating value. In contrast to the 1990s concern with consumerism, often equated with McDonaldization of society (Ritzer 2015), in this time of Global Brooklyn, everyone seems to be talking about "prosumers." A critical question here would concern the capture of value created in these social relationships by digital platforms for whom the traces of engagement around food become valuable data to be resold to companies.

Due to the centrality of the visual, there is little need to speak the same language, which allows Global Brooklyn to be adopted piecemeal, without reference to or embrace of discursive and ethical elements, even when some of these are visibly coded, as is the case with the aesthetics of hard work conveyed by clothing or reclaimed, battered stuff. The generative grammar that underlies the manifestations of Global Brooklyn around the world can be adopted with intensities and accuracy on a spectrum that goes from the wholesale acquisition of materials, practices, and symbolic elements—including values—to the use of the sensorium in and of itself, as a cool reference to interior design trends, but with limited consequences beyond that.

Food and food cultures are an excellent entryway to assess the dynamics of globalization.[2] Food scholars have provided different interpretations and perspectives on the global diffusion of products and practices. In the case of books focusing on single commodities—which have their most famous academic progenitor in Sidney Mintz's masterpiece *Sweetness and Power: The Place of Sugar in Modern History* (1985) and in the wildly popular *Cod: A Biography of the Fish that Changed the World,* by Mark Kurlanski (1998)—the approach can be both historical, following the development of product's uses and cultural meanings over time, and anthropological, adopting the perspective of the "social life of things," theorized by Arjun Appadurai and Igor Kopytoff (Appadurai

1986). Taking a cue from Kurlanski, in books meant for general readerships, the emphasis on a specific ingredient or product is often accompanied by outsized claims suggesting that the object of the book has somehow shaped or explains not only the culinary landscape but the economies and cultures of whole nations, if not the entire globe.[3] Global Brooklyn does not have such high aspirations.

Although the single product books for general readerships frequently deal with issues of colonialism and global capitalism, they rarely engage with underlying theory of globalization as a motivation for the choices of their content matter and organization. That is not the case when a product becomes a lens to gauge and critique value chains, trade flows, and economic power relations between the Global North and developing countries, within frameworks more or less explicitly influenced by Immanuel Wallerstein's (2004) world system analysis.[4] Stimulating anthropological research has explored worldwide food phenomena by connecting various locations with a place and a product that closely identifies with it.[5]

International collaborations have often been launched to deal with global phenomena.[6] Others have focused on diasporic communities to highlight and appraise globalization dynamics through food.[7] These elements are relevant to reflect on Global Brooklyn, as people, culinary culture, objects for food and drink preparation, as well as design elements seem to have influenced categories of taste among at least certain generations of consumers. Last but not least, research on the global expansion of transnational food corporations has exerted a great influence in shaping globalization ideas in food studies and the common use of the concept of "glocalization."[8] These studies highlight corporate policies and methods against the background of which Global Brooklyn stands apart as a noncentrally coordinated model of dispersion. There is no control center as in the case of large transnational corporations, but rather a recognizable set of design, sensory, and practice elements that have found fertile grounds in very different contexts.

The chapters and dispatches in this volume have precisely the goal of allowing a diffused production of knowledge and a more granular understanding of how what appears like a coherent global trend takes varied meanings and resonates with diverse cultural and social perspectives wherever it surfaces. For this reason, we argue that the elaboration of an ideal type provides enough hermeneutic flexibility and a point of departure for embedded and localized research, while refining shared categories. The ideal type approach can develop conversations that may shed a different light on how dispersed and embedded global dynamics and mechanisms shift and evolve over time and across space.

Ethnographers have long struggled to come to terms with understanding the interaction of global and local (Marcus 1995; Burawoy 2000; Appadurai

1996; Tsing 2005). Honed in extensive studies of one locale, very often imagined as disconnected and autonomous, the craft of ethnography had to be enhanced to capture the new global interconnectivities. One response has been to develop strategies of multisited research in which one scholar moves between locales, having developed an empirical or conceptual account of their linkage. This strategy significantly reduces the time for fieldwork in a given site. Another response, which we have adopted in this project, has been to attempt a collaborative ethnography, building and refining conceptual analysis from the bottom up by working with a network of researchers. The benefit of this approach is that not only the gathering of evidence but also theorizing happens bottom-up, drawing on localized debates and contexts. The resulting account of globalization appreciates both the world-scale political economy of digital platforms with its supply of new commodities, services, and narratives, and the localized processes of meaning-making and economic demands that make what we call "decentralized sameness" of Global Brooklyn so effective.

Decentralized Sameness

Global Brooklyn spaces tend to present themselves in opposition to corporate models where a singular total brand strategy centralizes control and small-scale entrepreneurs become franchisees, governed by multinational companies even when consumers see a cute couple starting a minuscule store. Global Brooklyn-type entrepreneurs aspire instead to autonomy in terms of aesthetic regimes, practices, and representations. In the most socially engaged cases, these can be expressions of the civic sensibilities that contest the disappearance of retail and family stores at the hand of large chains and corporations. In order to showcase independence, Global Brooklyn establishments are rarely located within shopping malls, even in cities where such venues are common and appreciated by most consumers. Typically found on street level and directly facing a sidewalk, they are literally "down to earth" in attaching themselves to pedestrian, street-food, improvisational urban imaginaries.

The emerging supply networks display proud independence from corporate structures (with the glaring and paradoxical exception of the cybercorporations that monopolize global platforms). Expressing a critique of contemporary food systems, the provenance of ingredients tends to reflect transparent and direct— if not personal—connections with producers or manufacturers, regardless of distance. Meat in butcher's shops or BBQ stores comes from trusted and possibly nearby animal farms, even as imported avocados are used as garnishes and cosmopolitan butchering blueprints guide the knife. Natural wines and craft beers are produced by frequently experimental but credible and committed artisans, even if they are not always dependable due to the sometimes makeshift nature of their operations. Global Brooklyn establishments often express what Michel de Certeau (1984) dubbed "tactics," that is adaptation, critique, and resistance within the territory dominated by strategies and decisions imposed from above, in this case a landscape marked by top-down corporate homogenization (except when dealing with the inevitable issues of credit, investment, and banking).

The lack of trust in grand narratives, regardless of whether they are commercialized through corporate practices or expressed through lofty and far-reaching political goals, leads to a focus on small-scale but tangible choices that introduce incremental changes toward health, sustainability, fairness, and justice. These forms of soft politics through entrepreneurial decisions express what designers Anthony Dunne and Fiona Raby (2013: 9) describe as a "shift away from the top-down mega-utopias dreamed up by an elite," caused by "dissatisfaction with existing models coupled with new forms of bottom-up democracy." However, it is legitimate to wonder—as many contributors to this volume will—whether Global Brooklyn has emerged to resist the productive needs of transnational capitalism or is rather just another reflection of its dynamics. It definitely is not the a-capitalist or anticapitalist approach it sometimes presumes to be; rather, as we have shown, it is aligned with many tenets of the current economy, including its capacity to coopt resistance.

Regardless of the social background and cultural proclivities of patrons and entrepreneurs, the overall result of the circulation of Global Brooklyn can be defined as "decentralized sameness." Wherever in the world they are located, Global Brooklyn establishments end up looking similar, whatever products and services they may offer. Such uniformity and familiarity, regardless of location and context, could be interpreted as a way to create a sense of comfort that is perceived as ethically, politically, and economically different from the "fake" ease offered by, say, McDonald's glocalized sameness.

Who Inhabits Global Brooklyn?

Global Brooklyn mimics industrial work environments and includes artisanal activities that partly belong to the sphere of the imaginary, as the working classes have increasingly embraced the convenience of mass-produced food.

The knowledge-intensive and self-aware performances of baristas, mixologists, butchers, and bakers, always ready to provide information and stories to clients, indicates a stylized and pastiche-driven return to materiality for postindustrial young millennials who may feel that crunching numbers and dealing with the virtual economy deprives them of a true contact with reality and social utility (Graeber 2018). Butchering courses, bread-making classes, and craft beer tastings are taken by people to constitute forms of soulcraft (Crawford 2009) and a new maker ethic, in which values and attitudes are entangled with hands-on practices and material objects, from ingredients to tools.

In North America and Western Europe, due to their resistance to corporate consumerism, Global Brooklyn inhabitants are sometimes identified with hipsters. In the most favorable interpretations, they are associated with an ever-shifting form of counterculture, always looking for original—and often frugal or DIY—expressions in taste and fashion; they want to resist mainstream consumerism, searching for individual styles all while sharing a recognizable collective identity (Nordby 2013). Their food consumption, often influenced by ethical and social concerns, tends to question commodification and manifests a preference for artisanal, farm-to-food, organic, and socially fair products. If mass-market brands are embraced, it is only with a sense of irony, nostalgia, or pastiche (Frank 1997; Lanham, Nicely, and Bechtel 2003; Cronin, McCarthy, and Collins 2014). At the same time, hipsters are often mocked as trend-obsessed, privileged middle-class posers looking to make themselves edgy and cool by hijacking and appropriating the objects, spaces, and practices of working class, ethnic, and immigrant communities. Furthermore, their looks and habits can be easily packaged for the mass market by commercial interests (Greif et al. 2010; Arsel and Thompson 2011). While participants in Global Brooklyn—both service providers and consumers—may be hipsters, they could also be foodies who build up their cultural capital through all things food, desire new experiences, and play an important role in processes of gentrification (Johnston and Bauman 2015; LeBesco and Naccarato 2015). However, outside of North America and Western Europe, Global Brooklyn as an aesthetic regime is connected with other social groups that would not necessarily be described as hipsters or foodies, including up-and-coming professionals and cosmopolitan upper-class nomads.

While Global Brooklyn attracts producers and consumers of all ages, at least in the Global North the majority of its participants are arguably gen X, gen Y (or millennials), and gen Z individuals. This generational composition may explain the fascination with certain characteristics of the Global Brooklyn aesthetic regime, such as industrial ruins, the analog, and old objects. These elements are newly connected with the North American and Western European middle-class millennial experience of lack of secure future, which is resentfully confronted as the legacy of older generations whose careers and perspectives constantly improved. The increasing rarity of secure, long-term jobs with pensions and the recent impact of economic downturns (including the consequences of the COVID-19 pandemic, which is taking place while we are editing this book) have curtailed opportunities for success and professional advancement for many millennials. Mobility has increased, and not just as leisure for the rich. Nomadic individuals, increasingly involved in the gig economy, have been forced to embrace remote work, establishing a different relation with employment and conventional offices. As new urban spaces for digital nomads in the form of coworking hubs à la WeWork proliferate globally, offering a promise of not only room but also community of like-minded freelancers, it is not an accident that their interiors are styled in the Global Brooklyn aesthetics.

In the Global North, educated, middle-class youth frequently do not feel they are winners in the new economy, as boomers and 1980s yuppies did. Perhaps they obsess about what they eat and drink in order to control the little they can in an otherwise increasingly uncontrollable reality. As they know they cannot conquer the world, their agency turns inward. Lifestyle becomes one last significant dimension in which to make choices, whereas the job market offers few. These shifts in the sense of future and agency cause intergenerational misunderstandings of what Global Brooklyn trends are and do. Coffee, elevated in terms of price and status, has become a symbol that older generations mobilize to vilify millennials and gen Xers as wasting money on overpriced flat whites and the maligned avocado toasts instead of becoming homeowners, as if cutting down on quality coffee could solve the global housing problem (Olen 2013).

Still, such concerns place Global Brooklyn in a temporal regime that Jens Beckert (2019) has described as the "exhausted future of neoliberalism," in which market-centric promises of growth and prosperity in the Global North have lost their capacity to legitimize the system, breeding discontent, populism, and a sense of malaise, but also a compensatory, if not decadent, interest in food. These perceptions often express themselves in the interest for sustainable economies and in the critique of progress as being founded on dehumanizing technology. On the ruins of growth-based civilization, practices emerge that reinvent the everyday along the lines of repair and maintenance (Mattern n.d.), the circular movement of matter, scaling down, and finding beauty in repurposed detritus, as a recent ethnography of drought-stricken Californians trying to make home among the anthropogenic climate crisis suggests (Vine

2018). Hence, the renewed interest for the analog and the physical connection with technology it allows, which assumes traits of quaintness as it is no longer connected with actual production activities. These aesthetic practices may also become ethical, because transforming materialities is sometimes an attempt at transforming selves. Even the most aestheticized of Global Brooklyn sites may contain seeds of reflection on contemporary personhood. Dunne and Raby (2013: 8) observed a certain "downgrading of dreams to hopes," because the twentieth-century dreams proved to be unsustainable and consumers need to readjust and perhaps scale down expectations. As a consequence, the new generations do not dream, they just hope they will not destroy themselves further. The experience of the exhausted future, as we have shown earlier, comes already wrapped in nostalgia.

These reflections are complicated when looking at the circulation of Global Brooklyn beyond the Global North. Postindistrialization is, after all, a Western story. China and India are going through fast development and modernization. Poland has been reindustrializing too. Millennials in India may not have expectations about security or a stable future. As a matter of fact, when they are employed in the most advanced sectors of the economy, they may enjoy more security than their predecessors. China's "little emperors," often only children, have access to larger resources than their parents with multiple siblings did. Educated Chinese youth can experience a sense of excitement and expect a brighter future than their peers elsewhere, despite political issues and the impact of crises such as the coronavirus epidemics. Global Brooklyn may mean something else in Vienna than in Shenzen, because the core of the world system has been shifting. In the wake of these transformations—discussed under the rubric of the new hegemony of China or the emerging markets of BRICS countries—there have been rearrangements in flows of capital, labor, and debt, leaving America's Y and Z generations in conflict with boomers and in a prolonged generational malaise that is unrecognizable in India or China. As a writer in the *New York Times* recalls,

> When I moved to New York City from Beijing in 2013, I had the surreal feeling that I had come from the future to a not-so-distant past. In China, I'd experienced six years of near relentless optimism. In the United States, I encountered a kind of pessimism and listlessness that I didn't expect. It was almost as if Americans had settled for a kind of Yelp-reviewed urban sameness: third-wave coffee shops and facsimiles of Brooklyn bars. The same is true in Europe. Same polished concrete floors, same nebulous sense of where you are. Purgatory is a café stacked with Kinfolk (Moxley 2019).

If this is the case, why are millennials in developing countries attracted to the postindustrial, exhausted future imaginary of Global Brooklyn? Is it just because it is cool, cosmopolitan, and allows them to participate in the experiences of what thy perceive as the epicenters of cultural and lifestyle trends? Or is it a ruse to brazenly display wealth and assert distinction by enjoying overpriced commodities consumed in a fashionable and fetishized manner? May there exist an Afrofuturist Global Brooklyn? Or is the appeal of Global Brooklyn in India, China, Brazil, or Turkey a form of resistance? Could the embrace of transnational aspirations be a nonpolitical reaction against the rise of local nationalisms and populisms that react to neoliberalism by violently closing borders and critiquing cosmopolitanism? It is surprising that no strong pushback against Global Brooklyn has developed even in the presence of powerful and growing antiglobalist feelings.

Similar conflicts and anxieties lurk beneath the apparent benign surface of Global Brooklyn, also in the Global North. Participants in its practices share a certain uneasy concern with issues of authenticity and cooptation. Are they participating in colonialist or classist dynamics? Are they selling out to ensure the success of their business or to enjoy respite from the hardships of the job market? Are they just hipsters who unintentionally support gentrification processes, refuse to admit their own hipsterism, and have a hard time dealing with the reactions of the displaced ones? Are they even aware of who the victims of gentrification are in terms of class, age, or ethnicity? How do they interact with spaces and movements of resistance to gentrification and affirmation of other cultures, whose actors may or may not be millennials?

How We Wrote the Book

These themes and concerns, from which we are not isolated ourselves, have provided the stimulus to look into Global Brooklyn. The first inkling that we were onto something developed during our fieldwork in Warsaw, Poland, to gather ethnographic material for a research project about the revaluation of local, regional, and traditional food in that country. As we were visiting bars, cafés, restaurants, stores, markets, and refurbished market halls turned into leisure spaces for trendy food consumption, we realized that many of those places looked and felt familiar. Not just designs and architectures, but the overall vibe of those locations reminded us of other places in different parts of the world that we had visited as researchers or tourists. We immediately—

and instinctively—came up with the moniker of "Global Brooklyn" to describe what we were observing. Later on, we would discover that the term had been used in a Facebook post by Lev Manovich, a media scholar, who in October 2015 posted a photo of a coffee shop in Riga with the hashtag #globalbrooklyn. Contacted for context, Manovich explained in an e-mail that he meant by that a "symbol for contemporary / urban / hip / creative—a combination of certain lifestyle and design trends + actual creatives living in the area" (Manovich, personal communication, 2018). After we included the term "Global Brooklyn" in an article about contemporary food in Poland on the *Huffington Post* (Parasecoli 2016), we started thinking more systematically about how and why cafés in Warsaw looked and felt similar to others in New York, in Mumbai, or Shanghai.

We took to posting pictures on Instagram with the hashtag #globalbrooklyn, involving our networks of colleagues and collaborators, our students, and whoever else was interested in creating a repository of images that could then be further explored. As we started talking with others about the idea and showing them the pictures, we were struck by how quickly our interlocutors understood what we were talking about, pointing out how many places like that they had patronized in their own city or while traveling. The concept was easy to grasp, especially from a visual point of view, although from the beginning we realized that underneath the apparent similarities there were also local differences that were worth exploring. As the available material grew, we began a more systematic reflection to explain such parallels among establishments in extremely diverse cultural and geographical contexts.

The essay that became the starting point for this volume appeared in *Food, Culture & Society* (Halawa and Parasecoli 2019). It was speculative and admittedly not an outcome of a structured and long-term research focused on Global Brooklyn itself. It was, rather, an invitation and a provocation we offered as we poured over field notes, images, and interview transcripts from other research into food cultures and lifestyles in various sites. Methodologically, it drew on cultural studies interest in the emergent sensibilities and desires coalescing into a new social formation (Williams 1978) and used a strategy of collaborative inductive coding of diverse data points in order to reveal patterns and stir up debate. In publishing it, we wanted to discuss and get feedback about the concept from other academics and students in the field of food studies. We felt we were starting a possibly long conversation: we wanted to research *with* Global Brooklyn rather than just research on it, also because we were very much part of that world.

Since we first presented our reflections at the 2017 annual meeting of the Association for the Study of Food and Society (ASFS), followed by the 2019 essay, we continued to expand our research through many talks in the United States, Denmark, China, Italy, and other countries, where we engaged audiences with the concept and invited them to contribute to our visual exploration by posting on social media and using the #globalbrooklyn hashtag. The small, growing repository of #globalbrooklyn images was imagined as a living visual archive allowing the development of new research questions and observations from around the world.

In developing this volume, we looked for contributors with diverse backgrounds, in order to include different approaches and frames of reference: among them are anthropologists, sociologists, media scholars, designers, social entrepreneurs, and authors with different styles and experiences. We distributed a draft of our original essay, with its ideal type and our reflections, and we asked the contributors to engage with it, push back against our analysis, and provide their own take, exploring what Global Brooklyn in their locations is and means. While we were working on this volume, we kept on presenting the research in conferences and talks as a work in progress, each time gathering feedback from our audiences and integrating it in our own analysis and in the conversations taking place with the contributors while they were drafting or revising their own chapters and dispatches.

We also asked the authors to pay attention to the role of the visual, often taking photos themselves, as the phenomenon is, as we have argued, partly constituted by the global online circulation of images. We created a shared repository of images that is for now available on Fabio Parasecoli's website fabioparasecoli.com.

The volume you are reading is organized into three parts. Part one presents the analytical toolkit, which the contributors mobilize and modify in their own explorations. Beyond this introductory chapter, it includes the first of what we call "dispatches," which are short-form impressions, narratives, or reflections that allow to further diversify the book in terms of themes, geographical distribution, and approaches. Dispatches enhance the more sustained arguments developed in the long-form chapters which, besides lively vignettes and more examples, offer broader reflections about cultural issues, social dynamics, and political tensions in the location the contributors engage with.

The second part collects chapters and dispatches from around the world, showcasing the variety of expressions of Global Brooklyn while focusing on the local particularities. For this reason, the chapters sometimes expand, sometimes push back on the very idea of Global Brooklyn as articulated in this introduction.

As we do not propose an Americanization thesis, in which a central location broadcasts a model abroad, each chapter provides elements for a complex theory of circulation, embeddedness, and variation of material and cultural elements. Contributors assess the relationship between the local and the global through various forms of mediation in terms of material objects, design, practices, and valuation processes. Furthermore, they explore dynamics of gentrification and the inherent tensions between a desire for ethos and community and the reality of upscale consumption. Overall, chapters and dispatches examine how the Global Brooklyn trend gets articulated in different locations and feeds back to other manifestations dispersed around the world, strengthening the cultural formation. They explore spaces as diverse as cafés, restaurants, wine markets, bakeries, coffee roasters, breweries, and street markets, with business models ranging from social entrepreneurship initiatives to the embrace of neoliberal priorities and goals.

The third and final part, entitled *Back to Brooklyn*, brings together all the suggestions, perspectives, and analyses provided by the contributors across chapters and dispatches. It also brings the insights from around the world to bear on two US locales: Chicago and, crucially and this time quite literally, Brooklyn itself. The conclusion returns to the introductory conceptualization of Global Brooklyn—not a stable and conclusive ideal type, but rather an evolving lens through which to investigate shifts in foodways—to reassess the concept itself and its analytic purchase in light of the contributions of our collaborators.

The #globalbrooklyn hashtag remains in use and we encourage readers to use it when stumbling upon establishments and practices that may remind them of what we have discussed in this book, engaging with its themes online.

Notes

1 Thought-provoking work has been developed on the diffusion of French cuisine and its rise as the pinnacle of global fine dining: Amy Trubek (2001) looked at the worldwide movement of professionals with specific skills and with a shared knowledge of food and drink, ingredients and dishes, as well as taste standards.

2 Following Dürrschmidt and Kautt (2019: 4–6), within food studies, we can identify different understandings of globalization: those inspired by the 1990s focus on hybridization, deterritorialization, and extreme mobility; those that from the 2000s question the inexorability of these processes through greater attention to the power relations that shape the complex relationships between the local and the global; and finally the more recent ones that look for more nuanced explanations of the mutual influences between the global circulation of people, goods, images, ideas,

and technology, inevitably embedded in very different contexts, each with their own rhythms, embodied experiences, as well as sensory and affective worlds.

3 Volumes have been dedicated to coffee (Morris 2019; Pendergrast 2010), tea (Moxham 2004; Rose 2010), spices (Freedman 2008; Schivelbush 1993), chocolate (Coe and Coe 2013; Moss 2009), curry (Collingham 2006; Sen 2009). As in the case of spices, the narrative can develop around specific nodal places or cities (Krondl 2007) or by following trade routes (Nabhan 2014).

4 Relevant work—both in monographs and edited books—has been conducted on green beans (Freidberg 2004), bananas (Striffler and Moberg 2003), broccoli (Fischer and Benson 2006), and coffee (Daviron and Ponte 2005).

5 Great examples are Theodore Bestor's work on sushi, fish, and the global network that extends from the Tsukiji market in Tokyo (Bestor 2004), George Solt's study on the global expansion of ramen (Solt 2014), and Anna Tsing's investigation of matsutake mushrooms in the US Pacific Northwest and its connections with Hmong mushroom pickers, traders and distributors, as well as Japanese gourmets (Tsing 2015).

6 It is the case of the collected volume by Richard Wilk and Livia Barbosa (2012) on rice and beans, which moves away from the "follow the food" genre to look at "how foodstuffs combine into meaningful dishes over long periods of time and large geographical spaces. . . . They show how 'sometimes foods move, and other times they stay in place, and become embedded in locality, ethnicity, nationalism, and other kinds of human groups'" (Wilk and Barbosa 2012: 4). By so doing, the contributors of the volume are able to examine slavery, nation-building, and neocolonialism.

7 A valid model is Richard Wilk's book on Belizean immigrants and the movement of their foodways between the Caribbean and North America, which presents ethnographic research together with the analysis of documents and media (Wilk 2006). A similar approach but with a stronger historical rather than anthropological inspiration can be found in Yong Chen's volume on Chinese food in America and Simone Cinotto's research on Italian immigrants in the United States (Chen 2014; Cinotto 2014).

8 We can mention James Watson's work on McDonald's in Asia (Watson 1997) and Amanda Ciafone's on Coca-Cola (2019).

References

Adkins, L. (2005), "The new economy, property and personhood," *Theory, Culture & Society*, 22 (1): 111–30.

Appadurai, A. (1986), "Introduction: Commodities and the politics of value." In A. Appadurai (ed.), *The Social Life of Things: Commodities in Cultural Perspective*. Cambridge: Cambridge University Press.

Appadurai, A. (1996), *Modernity At Large: Cultural Dimensions of Globalization*. Minneapolis: University of Minnesota Press.

Arsel, Z., and Thompson, C. J. (2011), "Demythologizing consumption practices: How consumers protect their field-dependent identity investments from devaluing marketplace myths," *Journal of Consumer Research*, 37 (5): 791–806. doi:10.1086/656389

Bartmanski, D., and Woodward, I. (2018), "Vinyl record: A cultural icon," *Consumption Markets & Culture*, 21 (2): 171–7.

Beckert, J. (2019), "The exhausted futures of neoliberalism: From promissory legitimacy to social anomy," *Journal of Cultural Economy*, https://doi.org/10.1080/17530350.20 19.1574867

Beer, D. (2019), *The Social Power of Algorithms*. London: Routledge.

Bestor, Th. C. (2004), *Tsukiji: The Fish Market at the Center of the World*. Berkeley: University of California Press.

Bucher, T. (2018), *If . . . Then: Algorithmic Power and Politics*. Oxford: Oxford University Press.

Burawoy, M., ed. (2000), *Global Ethnography: Forces, Connections, and Imaginations in a Postmodern World*. Berkeley: University of California Press.

Certeau, M. de (1984), *The Practice of Everyday Life*. Berkeley: University of California Press.

Chang, B. (2012), "Jennifer Mankins and the Brooklyn Aesthetic," *The New York Times*, December 26. Available online: https://www.nytimes.com/2012/12/27/fashion/je nnifer-mankins-and-the-brooklyn-aesthetic.html

Chayka, K. (2016), "Same old, same old: How the hipster aesthetic is taking over the world," *The Guardian*, August 6. Available online: https://www.theguardian.com/co mmentisfree/2016/aug/06/hipster-aesthetic-taking-over-world

Chen, Y. (2014), *Chop Suey, USA: The Story of Chinese Food in America*. New York: Columbia University Press.

Chumley, L. H. (2013), "Evaluation regimes and the qualia of quality," *Anthropological Theory*, 13 (1–2): 169–83.

Chumley, L. H., and Harkness, N. (2013), "Introduction: QUALIA," *Anthropological Theory*, 13 (1–2): 3–11.

Cinotto, S. (2014), *The Italian American Table: Food, Family, and Community in New York City*. Urbana: University of Illinois Press.

Clark, D. (2004), "The raw and the rotten: Punk cuisine," *Ethnology*, 43 (1): 19–31.

Cochoy, F. (2010), "How to build displays that sell," *Journal of Cultural Economy*, 3 (2): 299–315.

Coe, S., and Coe, M. (2013), *The True History of Chocolate*. London: Thames & Hudson.

Collingham, L. (2006), *Curry: A Tale of Cooks and Conquerors*. New York: Oxford University Press.

Couceiro, A. (2019), Elpais.com, September 9, 2019. Available online: https://elcomid ista.elpais.com/elcomidista/2019/07/23/articulo/1563895285_703028.html

Crawford, M. B. (2009), *Shop Class as Soulcraft: An Inquiry into the Value of Work*. New York: Penguin.

Cronin, J. M., McCarthy, M. B., and Collins, A. M. (2014), "Covert distinction: How hipsters practice food-based resistance strategies in the production of identity," *Consumption Markets & Culture*, 17 (1): 2–28.

Daviron, B., and Ponte, S. (2005), *The Coffee Paradox: Global Markets, Commodity Trade and the Elusive Promise of Development*. London and New York: Zed Books.

Deseran, S. (2013), "Coffee gone sour," *San Francisco Magazine*, November 26. Available online: https://issuu.com/saradeseran/docs/1410548491wpdm_15_sfmag_1213_c offee

Dunne, A., and Raby, F. (2013), *Speculative Everything: Design, Fiction, and Social Dreaming*. Cambridge, MA: The MIT Press.

Dürrschmidt, J., and Kautt, Y., eds. (2019), *Globalized Eating Cultures: Mediatization and Mediation*. Cham: Palgrave Macmillan.

Finn, S. M. (2017), *Discriminating Taste: How Class Anxiety Created the American Food Revolution*. New Brunswick: Rutgers University Press.

Fischer, E., and Benson, P. (2006), *Broccoli and Desire: Global Connections and Maya Struggles in Postwar Guatemala*. Stanford: Stanford University Press.

Frank, T. (1997), *The Conquest of Cool: Business Culture, Counterculture, and the Rise of Hip Consumerism*. Chicago: University of Chicago Press.

Freedman, P. (2008), *Out of the East: Spices and the Medieval Imagination*. New Haven and London: Yale University Press.

Freidberg, S. (2004), *French Beans and Food Scares: Culture and Commerce in an Anxious Age*. Oxford: Oxford University Press.

Graeber, D. (2018), *Bullshit Jobs: A Theory*. London: Allen Lane, an imprint of Penguin Books.

Greif, M. (2016), *Against Everything*. Place of publication not identified: Verso Books.

Greif, M., Lorentzen, Ch., Clayton, J., Pillifant, R., Horning, R., Baumgardner, J., Evans, P., Jefferson, M., Moor, R., Glazek, Ch., and Tortorici, D. (2010), *What Was the Hipster?: A Sociological Investigation*. New York: n+1 Foundation.

Halawa, M. (2011), "Nowe media i archiwizacja życia codziennego [New media and the archivization of everyday life]," *Kultura Współczesna*, 70 (4): 15.

Halawa, M., and Parasecoli, F. (2019), "Eating and drinking in global Brooklyn," *Food, Culture & Society*, 22 (4): 387–406.

Harkness, N. (2015), "The pragmatics of qualia in practice," *Annual Review of Anthropology*, 44 (1): 573–89.

Heuts, F., and Mol, A. (2013), "What is a good tomato? A case of valuing in practice," *Valuation Studies*, 1 (2): 125–46.

Ingold, T. (2013), *Making: Anthropology, Archaeology, Art and Architecture*. London: Routledge.

Irvin, C. (2016). "Constructing hybridized authenticities in the gourmet food truck scene," *Social Interaction*, 40 (1): 43–62.

Issenberg, S. (2014), *The Sushi Economy: Globalization and the Making of a Modern Delicacy*. New York: Gotham Books.

Jameson, F. (1991), *Postmodernism, or, the Cultural Logic of Late Capitalism*. Durham: Duke

Johnston, J., and Baumann, S. (2015), *Foodies: Democracy and Distinction in the Gourmet Foodscape*, second ed. New York: Routledge.

Jordan, J. A. (2015), *Edible Memory: The Lure of Heirloom Tomatoes and Other Forgotten Foods*. Chicago: The University of Chicago Press.

Krondl, M. (2007), *The Taste of Conquest: The Rise and Fall of the Three Great Cities of Spice*. New York: Ballantine Books.

Kurlanski, M. (1998), *Cod: A Biography of the Fish that Changed the World*. New York: Penguin Books.

Laclau, E., and Mouffe, Ch. (1985), *Hegemony and Socialist Strategy*. London and New York: Verso.

Lamont, M. (2012), "Toward a comparative sociology of valuation and evaluation," *Annual Review of Sociology*, 38 (1): 201–21.

Lanham, R., Nicely, B., and Bechtel, J. (2003), *The Hipster Handbook*. New York: Anchor.

LeBesco, K., and Naccarato, P. (2015), "Distinction, disdain, and gentrification: Hipsters, food people, and the ethnic other in Brooklyn, New York." In K. M. Fitzpatrick and D. Willis (eds.), *A Place-Based Perspective of Food in Society*, 121–39. New York: Palgrave Macmillan US.

Lewis, T. (2020), *Digital Food: From Paddock to Platform*. London: Bloomsbury Academic.

Marcus, G. E. (1995), "Ethnography in/of the world system: The emergence of multi-sited ethnography," *Annual Review of Anthropology*, 24: 95–117.

Mattern, S. (n.d.), "Maintenance and care," *Places Journal*, December 5, 2018. Available online: https://placesjournal.org/article/maintenance-and-care/?cn-reloaded=1

Mintz, S. (1985), *Sweetness and Power: The Place of Sugar in Modern History*. New York: Viking Penguin.

Morris, J. (2019), *Coffee: A Global History*. London: Reaktion Books.

Moss, S. (2009), *Chocolate: A Global History*. London: Reaktion Books.

Moxham, R. (2004), *Tea: Addiction, Exploitation, and Empire*. New York: Carroll & Graf Publishers.

Moxley, M. (2019), "Letter of Recommendation: Revolving Restaurants," *New York Times Magazine*, March 26. Available online: https://www.nytimes.com/2019/03/26/magazine/letter-of-recommendation-revolving-restaurants.html

Nabhan, G. P. (2014), *Cumin, Camels, and Caravans: A Spice Odyssey*. Berkeley: University of California Press.

Nordby, A. (2013), "What is the hipster?" *Spectrum*, 25: 52–64.

Ocejo, R. E. (2017), *Masters of Craft Old Jobs in the New Urban Economy*. Princeton: Princeton University Press.

Olen, H. (2013), *Pound Foolish: Exposing the Dark Side of the Personal Finance Industry*. New York: Penguin.

OpenBaladin. (2019), http://www.openbaladinroma.it/chi-siamo/?lang=en

Parasecoli, F. (2016), "Food in Poland: Much is changing, much remains the same," Huffingtonpost.com. Available online: https://www.huffingtonpost.com/fabio-par asecoli/food-in-poland-much-is-ch_b_12081760.html

Parasecoli, F. (2017), *Knowing Where It Comes from: Labeling Traditional Foods to Compete in a Global Market*. Iowa City: University of Iowa Press.

Parasecoli, F., and Halawa, M. (2018), "Reinventing Polish food as heritage and national identity," conference paper, *L'alimentation comme patrimoine culturel: enjeux, processus and perspectives*, Tours, France, November 15–17.

Pendergrast, M. (2010), *Uncommon Grounds: The History of Coffee and How It Transformed Our World*. New York: Basic Books.

Penin, L., (2018), *An Introduction to Service Design: Designing the Invisible*. London: Bloomsbury.

Quito, A. (2016), "Designers have an 8-letter word for the despised hipster aesthetic colonizing the planet," QZ.com, December 2. Available online: https://qz.com/76485 3/the-brooklyn-look-is-sweeping-the-world-to-designers-dismay-and-business-owners-delight/

Ritzer, G. (2015), *The McDonaldization of Society*, 8th ed. Los Angeles: Sage.

Rose, S. (2010), *For all the Tea in China: How England Stole the World's Favorite Drink and Changed History*. New York: Penguin Books.

Rousseau, S. (2012), *Food Media: Celebrity Chefs and the Politics of Everyday Interference*. Oxford: Berg.

Schivelbush, W. (1993), *Tastes of Paradise: A Social History of Spices, Stimulants, and Intoxicants*. New York: Vintage Books.

Schwarz, O. (2009), "Good young nostalgia camera phones and technologies of self among Israeli Youths," *Journal of Consumer Culture*, 9 (3): 348–76.

Sen, C. T. (2009), *Curry: A Global History*. London: Reaktion Books.

Sennett, R. (2009), *The Craftsman*. London: Penguin Books.

Solier, I. de. (2013), *Food and the Self: Consumption, Production and Material Culture*. London: Bloomsbury.

Solt, G. (2014), *The Untold History of Ramen: How Political Crisis in Japan Spawned a Global Food Craze*. Berkeley: University of California Press.

Striffler, S., and Moberg, M., eds. (2003), *Banana Wars: Power, Production, and History in the Americas*. Durham and London: Duke University Press.

Syvertsen, T., and Enli, G. (2019), "Digital detox: Media resistance and the promise of authenticity," *Convergence*. Available online: https://doi.org/10.1177/1354856519847325

Trubek, A. B. (2001), *Haute Cuisine: How the French Invented the Culinary Profession*. Philadelphia: University of Pennsylvania Press.

Tsing, A. L. (2005), *Friction: An Ethnography of Global Connection*. Princeton: Princeton University Press.

Tsing, A. L. (2015), *The Mushroom at the End of the World: On the Possibility of Life in Capitalist Ruins*. Princeton: Princeton University Press.

Usborne, S. (2014), "The spindly upper-case scrawl has been embraced to death by marketing folk to appeal to Young People and adorns everything from paperbacks to salad bags," *Independent*. Available online: https://www.independent.co.uk/life-style/fashion/features/how-did-hipster-sans-serif-become-the-defining-font-of-2014-9904544.html

Vine, M. (2018), "Learning to feel at home in the Anthropocene: From state of emergency to everyday experiments in California's historic drought," *American Ethnologist*, 45 (3): 405–16.

Wallerstein, I. (2004), *World-Systems Analysis: An Introduction*. Durham and London: Duke University Press.

Watson, J. L., ed. (1997), *Golden Arches East: McDonald's in East Asia*. Stanford: Stanford University Press.

Weber, M. (1997), *The Methodology of the Social Sciences (1903–17)*, translated and edited by E. A. Shils and H. A. Finch. New York: Free Press.

Wilk, R. (2006), *Home Cooking in the Global Village: Caribbean Food from Buccaneers to Ecotourists*. Oxford: Berg.

Wilk, R., and Barbosa, L. (2012), *Rice and Beans: A Unique Dish in a Hundred Places*. London and New York: Berg.

Williams, R. (1978), *Marxism and Literature*. Oxford: Oxford University Press.

Zuboff, S. (2019), *The Age of Surveillance Capitalism: The Fight for the Future at the New Frontier of Power*. London: Profile Books.

Zukin, S. (2009), *Naked City: The Death and Life of Authentic Urban Places*. New York: Oxford University Press.

Dispatch

Mobile Brooklyn: The Arrival of Food Trucks

Bryan W. Moe and Alexandra Forest

The years between 2008 and 2010 are when food trucks hit peak swag in Los Angeles, often considered one of the centers of the rise of a new generation of food trucks. Los Angelinos got on a hype-train over a Korean fusion taco sold by a tattooed street-food-sophisticate outside a bar in the Los Angeles neighborhood of Venice. While coffee shops and barbershop/brewery hybrids got wood and steampunk facelifts in Echo Park, CA, food trucks took elements of the Global Brooklyn aesthetic and welded it into V-8 retrofitted 15-foot lunch wagons. Taco trucks and California-dog carts have long been street food staples, but these new food trucks felt different, exciting, and controversial. So much so that any gastro-fusion hipster worth their Himalayan salt should have a trove of glowing Tweets and Yelp reviews for their favorite mobile kitchens.

Ten years have passed, and the food truck landscape has evolved in Los Angeles and elsewhere, becoming a global trend. However, even the V-8 steampunk food truck success is still linked to a design scheme that invited people to participate in public spaces in unique ways and at the right cultural moment. For example, early adoption of social and locative media using global positioning system (GPS) technology functioned to bridge gaps between customer and co-creator. More interaction, including in virtual spaces, led people to become emotionally attached to the chefs, their food, and the experience they were proposing, while inspiring others to search out the hype or fall victim to FOMO (Fear Of Missing Out). Furthermore, food trucks put on family-friendly, disco-type events that provided the community with public spaces in which to engage in dialogues ranging from what tastes good to more pressing issues of shared values, social justice, and environmental sustainability. In this short dispatch, we discuss the food trucker's connection to the Global Brooklyn aesthetic by exploring how it was used in five key areas: material design, renewed relationship with locality, communication, reflexivity, and renewed appreciation of manual labor.

Material Design

The design of the food trucks fit within the Global Brooklyn's hipster and postindustrial feel, as illustrated in this volume's Introduction, but unlike their brick-and-mortar counterparts, everything is less polished because it is constantly exposed to the elements. The first popular food trucker, Roy Choi, embraced that appeal. The myth goes, in late 2008 in Los Angeles Choi, then down and out chef, took a semiretrofitted taco truck with simple white siding, a mosaic of die cut stickers, and fresh street tagging to late-night crowds in need of tasty-alcoholic-absorption (Choi, Nguyen, and Phan 2013). The fluorescent light shone out of the motorized kitchen, providing a beacon for awaiting grub and virtue signaling selfies. Other early food trucks also altered the traditional taco truck aesthetic by slapping on colorful sticker-wraps, removing the taco truck staple brash neon signs, while adding soft track lighting and artfully crafted menus on chalkboards. More adventurous food truckers, like the owners of Maximus Minus in Seattle, made extensive modifications to their mobile kitchens by retrofitting the entire outside into a riveted metal war pig, complete with a front end that doubles as the war pig's snout and ears protruding from the roof (Maximus-Minimus 2019).

Renewed Relationship with Locality

The employees, chefs, and cooks are also part of the design of Mobile Brooklyn. In the case of Choi, it was unique to see a heavily tattooed Korean-American slinging Korean-Mexican fusion grub out of a taco truck. But his ethos was a central element in the appeal of food trucks and its connection to locality. Los Angeles is deeply influenced culturally by Hispanic and Asian communities, while being drenched in Hollywoodness and stuck in endless traffic. Choi can be seen as an authentic embodiment of this multiculturalism with pop-culture cachet. By no means is he the only figurehead, but rather a representation of the many sons and daughters of Los Angeles cooking on the streets. Through his highly visible profile, there was recognition and validation to others that the "locality" of the mobile kitchens, even if it could be perceived as a mishmash of influences, was an important feature of their success. Choi has also been quick to point out that his mostly Hispanic and female crew on the Kogi food trucks should be recognized on par with himself as key contributors to its continued success in that they are the daily face of the business and provide culinary skills by constantly updating and altering the menu.

Furthermore, the experiential factors of finding the truck, standing in line, and remaking the public space become key elements of co-creation through the connection between cooks and their clients. The "public" aspect frames social cues that invite to use one's voice with confidence or pick up a conversation with a stranger. The act of remaking social space also forces the customer to imagine new ways of interacting with the environment around them. A lack of or limited number of conventional tables and chairs present the audience with the task of reimagining the concrete parking header as a knockoff Van Doesburg concrete chair and side table. This approach goes farther in remaking entire unused parking lots, alleyways, and street corners into attractive outdoor cafés and marketplaces.

Communication

Food trucks may not have succeeded without the introduction of GPS and social media, specifically Twitter and Facebook. Early food truckers needed them to direct customers to their ever-changing locations. But they were also crucial because customers were very open to sharing their experiences and creating online and underground buzz. Suddenly, eating on the street in makeshift surroundings was "Instagramable." But like all good things, the new trend was swiftly picked up by more traditional media companies, which made the story of the food trucks viral and turned them into a great source of cultural capital among the general population.

Besides the large amount of publicity, social media gave food truck owners constant contact with customers. Food truckers communicating with their clients in this fashion functioned as a co-creative endeavor. Customers could give reviews directly to the person cooking the food and would often get responses from the chefs, both offline and online. In other words, these verbal exchanges were and still are an open forum. Before long networks of organizations were created, often connected to each other, from the Street Vendor Project in New York City to the LA Street Vendor Campaign, the National Food Truck Association, and other advocacy groups across the country that dealt with issues of labor and regulations.

Reflexivity

There is generally a lower barrier to entry into the street food world compared to brick-and- mortar establishments. Historically lower monetary investments, less training, and fewer skills were needed to become street food purveyors. But when professionally trained chefs like Roy Choi or Chef Crystal De Luna-Bogan of the Grilled Cheeserie truck in Nashville started cooking in mobile kitchens, the way the public viewed street food overall changed. Choi and others embraced a work ethos that took training seriously, helping a wider and larger audience establish clearer expectations and standards for street food presentation and quality. In so many words, the new approach gave an impression of professionality and safety with a dash of "cool."

Food truck customers were also more knowledgeable about what they were eating, who was cooking it, and how it was cooked. This falls in line with the general notion that since the early 1990s American audiences have become more knowledgeable about food and foodways based on an ever-increasing amount of food media consumed (Rousseau 2012). While this may not translate into actual cooking skills, it does educate consumers about different sets of ideals surrounding food and its role in self-identification. Food truck customer's ideals trend toward sustainable, organic, local, and gourmet food. Correctly or not, they have given preeminence to trained chefs or highly educated cooks. Unfortunately, the ability of these chefs to access and activate education creates a discrepancy in social status with traditional street food vendors that may have more experience and knowledge of local and traditional street foods.

Manual Labor

Thanks to their open design, most food trucks allow customers to see their food being prepared and watch as cooks "do work." For long periods in history, street food cooks were overlooked for their skills or negatively stereotyped by local governments and media (Gutman 1993). The rise of food trucks also shed light on more traditional street food cooks and their mobile kitchens and stalls. Loncheras became just as popular as some of the new, fancier food trucks. Los Angeles writer Jonathan Gold spent the last years of his life bringing shine to these communities. He claimed that some of the best food in Los Angeles was from the streets and often found in mobile kitchens. In fact, it was a fish taco truck, Mariscos Jalisco, that was a favorite of Gold's for years until his death (Gold 2014). While it is still odd that we needed someone like Gold or Choi to forge a connection and bring attention to these pre-food-truck, pre-Global-Brooklyn street food vendors, at least these celebrities were able to bring back some agency to them. This connection also symbolically generates prestige for current food truckers, as they are supposed to be sharing the same spirit and attitude, which of course is meant to make them more authentic (please read the hipster sarcasm, and then virtue signaling in this statement).

Conclusion

During the 2000s, marred by recession and social upheaval in the United States and globally, mobile kitchens became more valuable and important to the economic and cultural outlook of cities, while appearing increasingly designed and Brooklynified (Halawa and Parasecoli 2019). From the start, customers could see the cooking space through large windows rather than being shielded from the production of their food. The grit from the street provided a sense of authenticity. Those cooking the food were puzzling at first but later became significant cultural figures that felt more like friends. Through the digital connection between the truckers and their customers, "shared" experiences could be publicly capitalized upon and used to stoke virtual fires. In those online spaces, communities were built not only among customers but also among truckers. Support systems allow individual owners to seek the help they need to grow their business and achieve success in their various communities. Global Brooklyn aesthetics and cultural approaches run thick within the grease traps of Mobile Brooklyn and they are in part responsible for the longevity of food trucks. What was once considered a simple fad by some is now a stylish part of the lives of people not only in Los Angeles but also in many cities around the world.

References

Choi, R., T. Nguyen, and N. Phan (2013), *L.A. Son: My Life, My City, My Food*. New York: HarperCollins.

Gold, J. (2014), "This must be the place," *Lucky Peach Magazine*, Winter, 10–11.

Gutman, R. J. S. (1993), *American Diner: Then and Now*. New York: Harper Perennial.

Halawa, M., and Parasecoli, F. (2019), "Eating and drinking in Global Brooklyn," *Food, Culture & Society*, 22 (4): 387–406.

Maximus-Minimus (2019), "Maximus-Minimus." Available online: http://maximus-minimus.com

Rousseau, S. (2012), *Food Media: Celebrity Chefs and the Politics of Everyday Interference*. New York: Bloomsbury.

Part II

Exploring Global Brooklyn

1

Cape Town

Postindustrial Chic in a Changing Society

Signe Rousseau

Introduction

With its iconic Table Mountain, magnificent beaches, world-renowned wines, and an ever-expanding Waterfront development featuring stores that may seem more at home on New York's 5th Avenue (Jimmy Choo, Louis Vuitton, Montblanc, and others—often at extremely favorable exchange rates, for international visitors), Cape Town is often described as a "cosmopolitan" destination at the tip of Africa. Yet its role in the history of the Apartheid struggle is equally foregrounded in the global imagination in a way that serves both the local economy and presumably edifies the experience of international visitors. After shopping and lunch at the Waterfront, vacationers can take a ferry to Robben Island to visit the claustrophobic cell where Nelson Mandela and other stalwarts of the Apartheid struggle spent up to decades of their lives. Or they can book a walking tour through the Bo-Kaap, an inner-city neighborhood populated by (now) brightly painted houses: a latter-day expression of freedom from the formerly mandated exclusively white houses, once occupied by Malay slaves, the roads between which are now regularly obstructed by busloads of tourists snapping pictures of the locals. Cooking classes in private homes, thriving corner shops stocked with hard-to-find spices for biryanis, and a family-run shoe repair shop in operation for almost a century make the site a symbolic "phoenix" rising from hardships endured and conquered despite South Africa's ugly, troubled past.

But Cape Town is of course more complex—more wonderful, and typically more infuriating—than any simple juxtaposition between the extremes of Waterfront jaunts and Bo-Kaap tours could capture (even recognizing their potential similarities in terms of commodifying a range of experiences, largely aimed at overseas visitors with foreign currency to spend). With a diverse population informed both by a history of rebellion and by a desire to compete economically and culturally on the world stage, it is perhaps unsurprising that pockets of Cape Town also provide excellent examples of Global Brooklyn thanks to their wholesale absorption of hipster culture (staffed by mostly white, bearded males, sporting visible ink) in areas where that constructed ambience promises to be lucrative, such as previously disadvantaged neighborhoods experiencing rapid gentrification. In such areas, the aesthetic of postindustrial chic is particularly evident in new or reclaimed architecture featuring bare brick walls and Edison light bulbs, as is the number of bearded baristas and sommeliers in bespoke coffee shops and restaurants that occupy such spaces, including a few that regularly appear on the World's 50 Best list. Among these, The Test Kitchen, ranked no. 44 in 2019, is located in a complex that previously housed a flour factory, but which now hosts one of the preferred Saturday morning "neighbourhood markets" for upwardly mobile millennials of all races. Dinner at The Test Kitchen costs more than the average monthly salary of the majority of people living in South Africa, a country where the Gini-coefficient is regularly reported to reflect the highest inequality in the world (World Bank 2019).

In short, Cape Town provides a fascinating case study of the local manifestation of a globally fetishized material culture, discourse, representation, and set of values that the chapters in this book explore as Global Brooklyn. Yet the anchoring of these values in a unique geopolitical situation can also exacerbate how fundamentally at odds that aesthetic remains with the lived reality of a country still struggling to assert its place on a global stage while also respecting—and, sadly, often disrespecting—its history. While Cape Town is the focus of this chapter, remarks about the recent purported "overtaking" of that city by Johannesburg in terms of eating out and "foodie" culture are instructive of a trend that is not restricted to one city: "foraging and nose-to-tail eating might be fashionable—but they are nothing new. That's just called lunch in Soweto" (Twigg 2018). Similarly, lunch at a range of establishments in Cape Town may just as well be lunch in Brooklyn—with the exception of the actual neighborhood you have to step out into to get back to your car, where you will undoubtedly be faced with homeless people (of all races, genders, and ages) waiting to approach for a handout. Most patrons of such establishments will claim they "have nothing" to give, despite just having spent a small fortune on lunch. It is a spoken fiction so practiced among middle to upper-class South Africans that it rolls off the tongue as unthinkingly as replying "fine, thank you" when asked how one is: a thoroughly class-based and deeply ingrained platitude (notwithstanding that the issue of whether giving anything to homeless people is a complicated one; see, for example, Andrews 2014). So, while Cape Town's Global Brooklyn scene shares some elements with the growing number of "Brooklynized" areas around the world, there are a number of stark differences that persist as reminders that while you can bring the look and feel of what Brooklyn represents to Cape Town (or at least to certain establishments, and to those who can afford them), you cannot take South Africa out of Cape Town.

Neoliberalism and Middle-Classing

With some creative license, where embedded capital can manifest in any form—the Waterfront (officially known as the Victoria and Albert Waterfront, first established as a working harbor in 1860, later developed into a thriving commercial zone in the late twentieth century, and now boasting Africa's "greenest" shopping complex) represents the Manhattan of Cape Town. In that scenario, what is known as the "East City Precinct" (East City) would be its natural Brooklyn—at least insofar as it resonates with the idea of that neighborhood's "revitalisation by a new generation of 'authentic' hipsters in the borough that had always sought to define itself as different from—and superior to—Manhattan" (Lebesco and Naccarato 2015: 124).

Cape Town's East City is not the only location where the material culture of Global Brooklyn manifests itself. Edison light bulbs abound in restaurants around the city, as do menus featuring only local craft beers rather than any of the mass-produced offerings from industry giants like South African Breweries (SAB). This author recently tried the first rice lager on the local market at a tiny Momofuku-style restaurant in another part of town. The name of the brew, "My China"—local slang for *bud*, or *dude*—invokes the "insider" camaraderie typical of the collective identity of the mostly hipster community that inhabits and patronizes Global Brooklyn. But the East City is arguably the most *saturated* setting in terms of a concentration of establishments visibly arching—or searching—for an "anti-Manhattan" feel in Cape Town. In 2017, the area was described on a leading real estate site as having been, for many years, "regarded as the less desirable fringe of Cape Town's CBD [Central Business District], largely overlooked by the developers who have been reshaping the cityscape during the last decade. However, this has begun to change with recent upgrades attracting new businesses and injecting new life into the once run-down area" (Property24 2017).

This part of the city being "less desirable" for commercial development for many years stems in part from its proximity to District Six, a historically multiracial residential suburb bordering the CBD, which in 1966 then president P. W. Botha declared a "White Area" under the 1950 Group Areas Act (designed to segregate South Africa by race under the Apartheid regime). As a result, most of the 60,000 residents of the suburb were forcibly displaced to (now notoriously gang-ridden) townships in the "Cape Flats," over the next two decades, their homes and businesses destroyed. This large, flat area some twenty or so kilometers out of the city, is now characterized by makeshift shacks, many of which, in 2020, finally have electricity, running water, and satellite dishes. In 1979, following the establishment of a "White" Technical University in the area, a variety of people "consisting of religious groups and community figures established a community group calling themselves the 'Friends of District Six.'" They promoted the view of the area as "'tainted' land [which] ensured the failure of the Cape Town Municipality to re-develop a large part of the land. In 1987 the 'Hands off District Six (HODS)' alliance was established, aimed at preventing the redevelopment of District Six" (SAHO 2016). In 2019, fifteen years into South Africa's democracy, the process of reconciliation and reclamation of District Six remains complicated and incomplete, but the East City thrives in spite of—or *because of*—its "fringe-ness." Indeed, the businesses (or rather, business owners) that prosper here are good examples of what Deborah James (2019: 33) calls the "new middle class" of South Africa—a group that values "the positive qualities that work can endow, a capitalist-style interest in property investment, an inculcation of rational subdivision and allocation of one's income to diverse ends, and a redistributive approach to economic arrangements." In one interpretation, this growing middle class represents a version of neoliberalism which has not yet reached the "exhaustion" discussed in the introduction to this volume (see also Beckert 2019), but new economic means have certainly brought with them a decadent interest in food.

Eastbound and Down

It does not require anything as dramatic as crossing a bridge over New York's East River to reach the East City in Cape Town. A mere ten-minute drive from the Waterfront, one route will take you past the Houses of Parliament. Then you would drive past the Kimberley Hotel (the oldest in the city, where a sign on an outside wall proudly advertises "R10 [approximately $0.70 in January 2020] Jägermeister shots—all day, every day, the best deal in the world!"), and the Book Lounge, a small, independent bookstore that hosts launches for most politically progressive books published in the country, so if you drive by after 6 p.m. on a book launch night, throngs of people standing on the street clutching e-cigarettes and glasses of "book launch" wine are not an unusual sight. Do not go as far as Mavericks, a "Gentleman's Club" where a significant part of the entertainment is provided by women of Eastern European descent with questionable immigration statuses, which are anecdotally overlooked by the officials who staff the Cape Town Home Affairs offices around the corner, where every day hundreds of asylum seekers stand in queues waiting for papers to confirm a new life in South Africa. Just before Mavericks, you'll find the flagship store of Truth Coffee Roasting.

Truth is arguably the epicenter of Global Brooklyn in Cape Town, particularly as the trend is framed by reporter Preeti Varathan in a 2018 *Quartz* video widely shared on social media. According to the report, the look—perhaps the "first truly global aesthetic"—began with Starbucks ostensibly teaching the world that "coffee wasn't just fuel; that coffee shops could be a place where people eat, and drink, and slow down. Starbucks made coffee a luxury experience. And as Starbucks went global, it spread that expectation around the world" (Varathan 2018). The US financial crisis, however, turned attention (back) to the "super lo-fi," "Brooklyn, hipster, nostalgic look," writer Kyle Chayka told Varathan, followed closely by Silicon Valley and Instagram, which essentially "homogenizes aspiration—everyone aspires to towards [*sic*] the same symbols of luxury at the same time." As the vogue spreads on social media, it gets "refined down to its most Instagrammable elements—a nostalgic, stripped back, luxury minimalism. . . . Those digital spaces are really amenable to the minimalist aesthetic, because the Web is kind of empty and blank anyway" (Varathan 2018). Varathan concludes the report with the assertion that "That's how coffee shops around the world came to look the same."

The "look" referred to above is indeed crucial, but it should be noted at this point that, while Edison bulbs and bare brick walls do exist as common denominators in many of the establishments discussed here, "look[ing] the same" is less relevant to the Cape Town Global Brooklyn aesthetic than a more intangible sense of belonging to, or at least identifying with, a particular community.[1] The group that Cape Town's Global Brooklynites identify with is indeed more international than it is local, particularly given the deep cultural, economic, and (still) to some extent, racial trenches that continue to divide the country. It is a group that enjoys the conceit and privilege of distinguishing itself in a young and emerging local market—and one which is by no means free from the shadows of various forms of colonialism that defined it, and its place in the world, for longer than most people can remember.

It is therefore unsurprising that Truth does not, in fact, look like just any coffee shop around the world. It is based on a "steampunk" aesthetic, explained by owner David Donde (2013) as "science fiction from a Victorian person's point of view," where "things that look found are made, [and] things that look made are found," as he described his shop after it was named the Best Coffee Shop in the World by *MSN Travel*'s Tom Midlane in 2013. Dominating the bare brick, "reclaimed" industrial feel of the space is an enormous machine specially designed to capture (and manufacture) the "essence of coffee." It is, in Donde's words, "not a coffee dispenser; it's a collection of chains and stored mechanical energy, and ratchets and barrels and lights, and you wind this thing up, it pings when it's ready, you kick a lever in, a series of gates opens and closes, beans fall ... it's kinetic sculpture, is what it is" (Donde 2013). It is also, in the context of the country's history, no less a monument to excess than the Waterfront complex, and no more welcoming to the sizeable portion of the population on its doorstep for whom ZAR32 for a shot of espresso (or ZAR40—approx. $2.70—for Single Origin Burundi blend) is a prohibitive price.

Still, Truth's brand is a perfect example of what the late *L.A. Times* food critic Jonathan Gold (2008) called the "third wave of coffee connoisseurship," where "roasting is about bringing out rather than incinerating the unique characteristic of each bean, and the flavour is clean and hard and pure." The success and presence of its brand on the global market was consolidated by being voted no. 1 in *The Telegraph*'s poll of Best Coffee Shops in the World in 2015 and 2016. Yet Truth was not the first thirdwave coffee shop in Cape Town—indeed, owner Donde originally made his mark on the city when he and erstwhile business partner Joel Singer opened Origin Coffee Roasting in 2006 in De Waterkant (literally "the water's edge," an area much closer to the Waterfront, and one with its own story of rapid gentrification long before the East City's). True to the designation of the "third wave" coffee enthusiasm that stalks Global Brooklyn, both establishments now host Barista Academies and can lay claim to producing some of the city's most well-calibrated microfoam.

Origin Story

South Africa had to wait a little longer than the United States for its own "Starbucks moment" (referring simply to the introduction—and subsequent expectation—of good coffee made fast, as mentioned above). The actual chain did not launch here until 2016 to much initial fanfare, as is often the case when an iconic US brand lands in the Global South. For example, the queues for the first Burger King in 2013 opening were ridiculous, as were those for the more recently opened Cape Town outlet for Krispy Kreme donuts, confirming many anecdotal observations that what happens in the United States, particularly as depicted on the big screen, has a notable cachet within and influence on a particular demographic of this country. Look no further than university students wearing oversized jeans below their hips, "hip-hop-style," not to mention the bearded, inked baristas that form part of the topic of this chapter.

But Cape Town-based coffee chain Vida e caffè (Portuguese for "life and coffee," colloquially known just as Vida) opened in 2001 and is now omnipresent not only in major hubs across the country but also in gas stations both in and out of cities, fuelling drivers across the nation. In 2009, editor-in-chief of *Monocle*, Tyler Brûlé, reported on Vida's arrival on "European shores," quipping that with its "sunny, southern hemisphere disposition," perhaps the concept is "just another Starbucks in giraffe's clothing" (Brûlé 2009).

After Vida introduced Cape Town to good coffee made fast (and loudly—a staff rule seems to be that baristas repeat all orders by shouting them to each other), Singer and Donde's Origin was one of the city's first tastes of "high quality coffee, poured by professionally [in-house] trained baristas using quality beans roasted with pride by a skilled artisan roaster" (Origin 2018). Origin's website claims to have trained "over 4,000 of the top baristas on the continent," and that the team "helped to found the Speciality Coffee Association of Southern Africa (SCASA) and to launch the South African National Barista Championships, affiliated with the World Barista Championships (WBC)." Origin HQ is not just a coffee shop specializing in beans from Africa (their offering includes beans from Ethiopia, Uganda, Kenya, and Rwanda); it is described as a "beautiful brick warehouse" (which indeed it is), its unique, coffee-bean colored floor achieved by having laboriously compressed volumes of coffee beans into a slick, smooth surface. So simple in design, so complicated in execution, much like the connotations of names like Truth and Origin (contrary to what is suggested by the name "The Art of Duplicity," a secretly located speakeasy also situated in the East City). These are not just coffee shops—they are carefully curated "experiences," and the main commodity on offer is "authenticity."

Not unlike its conceptual counterpart Manhattan, (Global) Brooklyn is a society of the spectacle, as Guy Debord titled his 1967 Situationist critique of consumer culture, in which he maintains that "the spectacle is not a collection of images. Rather, it is a social relation among people, mediated by images" (Debord 1996: 4).[2] Debord's critique was based on a (regrettable, to his mind) sense of passivity on the part of consumers, well captured by the image publisher Black and Red chose for the 1983 edition of the book—a photograph of the audience at the premiere of the first "natural vision" (3D) movie in 1952, transfixed by the screen, their 3D glasses giving them the illusion of immersion while no one in the room actually interacted with one another. Our present-day spectacle, by contrast, is more accurately defined by the constant *activity* of consumers, particularly when it comes to documenting experiences on social media. The need to provide fodder for such attention is among the challenges faced by any business, even—and slightly paradoxically—for one which strives to build an identity as inhabiting a fringe, providing its patrons the opportunity to do the same.

A location such as a Starbucks, a Vida, or a Burger King communicates sameness, predictability, and—usually—reliability (none of which are necessarily bad qualities), wrapped as they are in the familiar veneer of a ubiquitous brand. Locations such as Truth and Origin, on the other hand, tell a story, or at least signal that they have a story to tell. Similarly, the name Nude Foods, as the first plastic-free grocery store in Cape Town is called, is clearly intended to set it apart from the "wrapped," or "processed," and likely mass-produced, foods consumers would find in any other supermarket, thereby signaling its "otherness" from the mainstream.

Truth, Origin, and Nude might be hard-pressed to deliver an all-encompassing narrative to satisfy the choice of a single word with a host of potential interpretations. Which, or whose, truth? The origin of what? Just nude, or denuded? What of? (Do they sell ingredients that The Naked Chef would cook with?) But just as there cannot be a spectacle without spectators, storytellers need listeners. Or Instagrammers. And they are, now, legion; many on the lookout for something "authentic" to post to stand out from the virtual crowd that most of us inhabit.

As Leigh Chavez Bush (2019: 21) writes, in striving to keep up with the "digitization" of everything,

> Users not only want storytelling; they demand authenticity. . . . As a key source of material and social culture, food dovetails superbly with the digital intimacies crafted through Web 2.0. At the same time, as an immediate, fragmented, and hybridized media format (. . .) , Web 2.0 has complicated our sense of authenticity, a rather nebulous word saturating dialogue about food, consumer culture, media, and much of contemporary life.

This observation rings as true in South Africa as it does anywhere with a social media penetration (likely *everywhere* except, perhaps, North Korea), and the reminder of how the search for authenticity permeates "much of contemporary life" is pertinent, given the current scourge of "fake news"—a phrase often, and puzzlingly, presented in ironic quotes.

There is little irony in fake news designed to damage democracies and demonize already-marginalized groups of people. There is, however, potential irony in expecting industrial architecture, Edison bulbs, and superior latte art to promise a "safe space" from the ills of modern living (such as mass-production resulting in potential—and unnecessary—harms to livestock, agriculture, and health of consumers and economies, for example). Not because such spaces cannot signal a conscientious distancing from those ills. Nor because it evidently takes a particular interior design—or offer of an unwrapped lentil— to convey such safety. One account put it in the context of the then-imminent opening of Nude Foods in the East City in Cape Town, referring specifically to the elimination of excessive plastic wrapping, "The concept is not trendy but rather wholly necessary, and a fundamental step in the right direction in waste reduction" (McGeever 2017). This can be applied to any number of contemporary choices like preferring single-origin coffee in order to honor small-scale farmers over large, exploitative conglomerates. That article also echoes a *Guardian* piece on zero-waste supermarkets in the UK, which argues that while such a trend is being "sold as radical 21st-century eco-vanguard stuff, . . . the organizing principle is the same as that familiar to a generation who grew up before plastics: you bring your own jars, sacks and pots, weigh what you want and cart it home" (Haynes 2017). Closely related to the ethos of Nude Foods, the practice of single-use plastic is no less scrutinized in South Africa than in the UK. A recent study by Professor Peter Ryan at the University of Cape Town aimed to ascertain whether plastic in the ocean is chiefly the result of land or sea pollution, which involved the researcher and his team actively *adding* plastic to the sea in order to track its movement—and thereafter to remove it—following a similar study conducted in Asia (Ryan et al. 2019.)

Truth Laid Bare

There cannot—nor should there—be any disavowing the problems with many of our industrial food-production and delivery systems. The coverage of the United Nations Global Action Summit in New York in September 2019 (enlivened by the presence and participation of sixteen-year-old climate activist Greta Thunberg, whose piercing glare at President Trump has already been immortalized on social media) was yet another reminder that we *are* in the process of destroying the planet. So moves to minimize (or eliminate) plastic usage, food waste, overfishing, and the exploitation of farmers and agricultural land, to name some of the main culprits, should be a priority for any global citizen with a functional moral compass. If that is the ethos of Global Brooklyn, it should be applauded and welcomed everywhere (with the unrealistic caveat that its offerings could also be available to people of all classes, which in this country will remain a distant fantasy for some time yet).

But caring about the environment does not require Edison light bulbs. In fact, they may be the *least* environmental friendly choice available.[3] And stripping walls, maintaining a carefully groomed beard, or relying on the equivalent of a steam engine to produce a cup of coffee does not combat fake news, economic inequality, or racial injustice, just as Michael Pollan's injunction to not "eat anything your great-great-great grandmother wouldn't recognize as food" (Pollan 2006) is not a useful way to combat obesity or the presence of additives in the food you choose to feed yourself or your family. These are all legitimate choices in the marketplace of globalization, as is the choice to buy a burger from a joint that prefaces "junk food"—that most maligned of designations when it comes to why what we put in our mouths is making us sick—with the word "gourmet," rather than going to McDonald's. But the oxymoronic "gourmet junk food" may indeed be a way the Global Brooklyn community chooses to differentiate itself; a route for hipsters to "achieve proper representation of their collective identity within the marketplace. Mundane consumption emerges as motor-force in allowing these consumers to surreptitiously maintain distinction and to protect their within-group identity from mainstream co-optation," at least according to Cronin, McCarthy, and Collins's (2014: 2) work on the "production of identity" trends in the global hipster community.

For Cape Town, one jewel in the crown of the Global South, the availability of these choices make many of its residents richer: we can compete for Instagram attention with our twelve-hour smoked brisket or award-winning coffee alongside people *in* Brooklyn—as can business owners who achieve international recognition for their efforts in a previously maligned "fringe" area of the city. But just as any number of cocktails served with artisanal ice in Brooklyn is not going to eliminate the news (fake or not) a resident in that borough may have to confront the next morning, no amount of gourmet junk food, artisanal coffee, or (admittedly good) cocktails in the East City of Cape Town is going to prevent the possibility of stumbling onto a street crowded with people protesting gender-based violence outside the Houses of Parliament. Nor should the celebration of plastic-free food or the opportunity to join a Barista Academy obscure the fact that most of the people who grew up less than five minutes away were denied access to such opportunities by their nonpeaceful, and nonnegotiable, relocation from the area decades ago to places more than a fifteen-minute drive away, which often forced a life of crime and poverty rather than empowerment.

Cape Town's East City Precinct is just one of many spaces in a country lauded in the twenty-first century for its seemingly miraculous ascension to a "rainbow nation" (per then-archbishop Desmond Tutu's description of a mythical land of racial harmony in 1994, when Nelson Mandela was sworn in as the first president of the new democracy). The invitation extended by the East City Precinct and other pockets of the metropolis to experience that "Brooklyn-feel" plays an important role in the economic, cultural, and social fabric of Cape Town. For locals and visitors alike, it is another attraction in the fairground of globalization, creating opportunities for employment, creativity, and Instagram fame. In many ways, it is a colorful expression of how far the country has come. Yet its existence is also a reminder that the rainbow (latte) nation remains a myth.

Notes

1 There is a broader political argument, which there is little space to broach here, but which resonates with 2019 Hitchens Prize Winner George Packer—a staff writer for *The Atlantic*—who argues that writers (as just one example of the proliferating factions of the twenty-first century) these days are "expected to identify with a community and to write as its representatives. . . . Groups save us a lot of trouble by doing our thinking for us. Belonging is numerically codified by social media, with its likes, retweets, friends, and followers. . . . In the most successful cases, the cultivation of followers becomes an end in itself and takes the place of actual writing" (Packer 2020).

2 Situationism was a Marxist-inspired critique of commodity fetishism and the perceived brainwashing of consumers by hegemonic narratives. It grew out of a French movement concerned with what they termed "psychogeography," or the impact of surroundings (e.g., the design of a city) on people's behavior.

3 As a *New York Times* piece notes on the trend of using Edison—or "filament"—bulbs in restaurants, some "do not produce enough light to be included in the higher federal efficiency standards that begin taking effect in 2012, but can use roughly three times the energy of a standard incandescent" (Cardwell 2010).

References

Andrews, G. (2014), "Should you give to people on the street?," *Synapses*, May 12. Available online: https://www.synapses.co.za/give-people-street/ (accessed January 25, 2020).

Beckert, J. (2019). "The exhausted futures of neoliberalism. From promissory legitimacy to social anomy," *Journal of Cultural Economy*, 13 (3): 318–30.

Brûlé, T. (2009), "Monocle weekly," *Monocle*, June 3. Available online: https://www.bei ngtylerbrule.com/tag/monocle-weekly/ (accessed September 30, 2019).

Bush, L. C. (2019), "The new mediascape and contemporary American food culture," *Gastronomica: The Journal of Critical Food Studies*, 19 (2): 16–28.

Cardwell, D. (2010), "When out to dinner, don't count the Watts," *New York Times*, June 7. Available online: https://www.nytimes.com/2010/06/08/nyregion/08bulb.html (accessed October 30, 2019).

Cronin, J. M., McCarthy, M. B., and Collins, A. M. (2014), "Covert distinction: How hipsters practice food-based resistance strategies in the production of identity," *Consumption Markets and Culture*, 17 (1): 2–28.

Debord, G. (1996), *The Society of the Spectacle* [1967]. New York: Zone Books.

Donde, D. (2013), "Cape Town coffee shop world's best," *You*, October 18. Available online: https://www.news24.com/You/Archive/cape-town-coffee-shop-worlds-best-2 0170728 (accessed October 20, 2019).

Gold, J. (2008), "La Mill: The latest buzz," *L.A. Weekly*, March 12. Available online: https ://www.laweekly.com/la-mill-the-latest-buzz/ (accessed January 24, 2020).

Haynes, G. (2017), "Bulk buy: Why zero-waste supermarkets are the new, old way to shop," *The Guardian*, August 31. Available online: https://www.theguardian.com/li feandstyle/shortcuts/2017/aug/31/back-to-the-future-the-zero-waste-supermarket (accessed November 10, 2019).

James, D. (2019), "New subjectivities: Aspiration, prosperity and the new middle class," *African Studies*, 78: 33–50.

LeBesco, K., and Naccarato, P. (2015), "Distinction, disdain, and dentrification: Hipsters, food people, and the ethnic other in Brooklyn, New York." In K. M. Fitzpatrick and D. Willis (eds.), *A Place-Based Perspective of Food in Society*, 121–39. New York: Palgrave Macmillan.

McGeever, C. (2017), "We chat to the guy behind Cape Town's first plastic-free grocery store," *Food24*, October 12. Available online: https://www.food24.com/News-an d-Guides/Features/we-chat-to-the-guy-behind-cape-towns-first-plastic-free-grocer -20171012 (accessed October 30, 2019).

Origin Coffee Roasting. (2018), "About." Available online: https://originroasting.co.za/v 3/introduction/ (accessed January 10, 2020).

Packer, G. (2020), "The enemies of writing," *The Atlantic*, January 23. Available online: https://www.theatlantic.com/ideas/archive/2020/01/packer-hitchens/605365/ (accessed January 25, 2020).

Pollan, M. (2006), "Six rules for eating wisely," *Time*, June 11. Available online: http://content.time.com/time/magazine/article/0,9171,1200782,00.html (accessed February 6, 2020).

Property24. (2017), "Cape Town's East City Precinct now the 'place to be,'" *Property24*, November 30. Available online: https://www.property24.com/articles/cape-town s-east-city-precinct-now-the-place-to-be/26766 (accessed January 10, 2020).

Ryan, P., Dilley, D., Ronconi, R., and Connan, M. (2019), "Rapid increase in Asian bottles in the South Atlantic Ocean indicates major debris inputs from ships," *Proceedings of the National Academy of Sciences*, 116: 42.

SAHO. (2016), "District six is declared a 'White Area,'" *South African History Online*, February 28. Available online: https://www.sahistory.org.za/article/district-six-decla red-white-area (accessed November 25, 2019).

Twigg, M. (2018), "How Johannesburg became the foodie capital of South Africa," *The Independent*, February 26. Available online: https://www.independent.co.uk/trave l/africa/johannesburg-restaurants-foodie-holidays-eating-south-africa-marble-che -epicure-a8224976.html (accessed October 15, 2019).

Varathan, P. (2018), "Coffee shops around the world are starting to look the same," *Quartz*, October 25. Available online: https://qz.com/1436053/coffee-shops-arou nd-the-world-are-starting-to-look-the-same/ (accessed September 30, 2019).

World Bank. (2019), "South Africa overview," Updated March 2019. Available online: https://www.worldbank.org/en/country/southafrica/overview (accessed July 15, 2019).

2

Melbourne

Care, Ethics, and Social Enterprise
Meet Global Café Culture

Tania Lewis and Oliver Vodeb

Introduction

Melbourne has become a striking gastronomic metropolis, known for one of the world's most acclaimed coffee cultures and an artfully aestheticized, hipster-friendly café scene. The city is also known—at least among global sustainability and environmental circles—for its grassroots green politics and community-driven alternative food movement (Lewis 2015: 348), a set of concerns reflected in the look, feel, and ethics of many of its celebrated restaurants and cafés. This chapter seeks to critically examine the phenomenon of Global Brooklyn through an ethical and food activism lens. Discussing a wide range of Melbourne cafés, from socially and ethically aware establishments to those adopting social enterprise models, we reflect on the ways cafés increasingly seek to combine profit with notions of "care."

In doing this, we seek to enhance the political and ethical dimension of the Global Brooklyn heuristic and to speak to the notion of care—planetary, justice, environmental, and community. We are interested in examining what might be the politics at play behind the transnational dissemination of a certain kind of postindustrial, DIY "look" and feel? What sort of shifting and emergent economic models might adhere to cafés that, in many ways, represent themselves as coming out of a more post-materialist culture, as anti-brands of a sort? In this chapter, we are interested in opening up and expanding questions around the politics of Global Brooklyn through exploring some of Melbourne's café spaces. Some questions we want to ask include the following: Is the cultural diffusion of a cosmopolitan café culture purely a bourgeois form of what George Ritzer

(2009) famously terms "Mcdonaldization" or what the editors of this collection discuss in terms of "decentralized sameness"? Is the DIY, anti-brand aesthetic of these global cafés a brand in and of itself? Or are we seeing more complex cultural, political, and economic trends being played out? What might be the points of tension and disruption within what outwardly might appear to be a shared lifestyle aesthetic, essentially embedded in consumption?

In the context of Melbourne's cutting-edge café culture, the neighborhoods of Fitzroy and nearby Collingwood are perhaps the hipster heart of the city of Melbourne. Once an extremely poor area, where Victorian workers' cottages and a plethora of factories were built on swampy insect-infested land, these neighborhoods right next to Melbourne's Central Business District have been gentrifying over the past decade and are now a sought-after inner-urban locale. Tiny and dilapidated Victorian terraces are becoming converted into architecturally revamped designer abodes for an upwardly mobile creative class. Huge red brick factories that, in the nineteenth century, accommodated chocolate manufacturers and flour mills now house slick apartments, with ground street frontage sporting an array of top-notch cafés, such as one of Melbourne's most famous, Proud Mary. In Collingwood in particular (as Fitzroy becomes unaffordable for many start-ups and creatives) we see once industrial blocks now housing a hodgepodge of cutting-edge and grassroots enterprises such as Vice Media and not-for-profit organizations like Stephanie Alexander's Kitchen Garden Program, which sets up productive gardens in schools across Australia. These various organizations sit near myriad cafés, often merging with the built environment of workplaces and blurring the boundaries of commercial, public establishments and spaces of work. Co-working cafés are also common features of the Collingwood café mix, all of which tend to sport interiors crafted in postindustrial concrete and recycled wood.

Tucked on the side off Brunswick street (the main strip and tourist hub of Fitzroy), Café Louis is an elegant, minimalist place communicating self-confidence and sophistication. In the midst of an area with dozens of other cafés, Louis, which focuses on breakfast and brunch, distinguishes itself by its upper-class fine, chic atmosphere and culture. The interior is slick, its menus and visual identity even more so. Like many Melbourne cafés, Louis's main communication tool is Instagram, where it displays a mix of highly curated and aestheticized food photography, intimate moments, and aesthetic impressions. Its food is similarly visually and sensually rich; when we ordered the Superfood Salad—with spiced cauliflower, freekeh, cucumber, cherry tomato, cranberries, kale, sunflower seeds, pepitas, sweet potato crisps, parsley, mint, and agave dressing—

the dish suggested an overwhelming sense of sophistication and promise of pleasure. Within Melbourne's extensive range of café offerings, Café Louis lies at the highly polished end of the spectrum, providing a dining experience that is both stylish and somewhat generic. Café Louis's owners both hail from India. While sitting in the quietly comfortable café space surrounded by well dressed, white middle-class customers, one gets, however, little sense of the diversity and vibrancy of the neighborhood and city just beyond its doors. In Café Louis one could indeed be in any high-end café in Europe or North America.

Right next door to Louis whose hidden door and minimal signage are easy to miss for the customer not "in the know," we find a very different café experience in the form of the hipster hangout Grub. Grub is open to the street and fronted by a 1965 Airstream van, which is parked in a wild garden-courtyard bounded by graffitied walls and Melbourne's distinctive nineteenth-century bluestone lanes. In contrast to Louis's coolly select interiors, Grub openly welcomes passersby into its large café space—a mixture between a glasshouse and a student share house, complete with a ping-pong table, lounge chairs, and a variety of plants including fruit trees and vegetables. While slightly shabby and bohemian in appearance, Grub's food is perhaps even more sophisticated than Louis's. Though self-described as "simple, good" food, the highly cosmopolitan menu carefully blends elements of Asian and European cuisines with locally sourced ingredients. The result is restaurant-quality food served in a deliberately and self-consciously laid-back environment. The look here is perhaps not so much the bare wood and concrete that characterizes so much of Melbourne café design. It does, however, share a similar concern with a DIY look and feel, with vintage tables and chairs, and the green plant-laden aesthetic that in Melbourne tends to be a marker of organic cafés offering locally sourced food. Such markers of taste often go hand in hand with a more socially and environmentally oriented enterprise model. We are not surprised to find that next to Grub's unisex toilets there are posters critiquing Australia's racist policies around national identity and migrants.

Our third example, café Streat (which we discuss in more detail below) has built political concerns centrally into its social entrepreneurship model, which at its core is about training and supporting marginalized, often homeless young people through its various cafés and its coffee roastery and artisan bakery. Its Collingwood-based center, where it houses its baking and coffee-roasting activities, is also home to a vibrant café, which, like Louis and Grub, offers the customer highly designed and aestheticized experiences. Worthy of being featured in an interior design magazine, the café is located in a repurposed 1860s manor. While the exterior still looks like a formal nineteenth-century house, its interior features a warehouse-style, industrial aesthetic with recycled materials, plywood walls, and a large amount of greenery hanging from beams. Streat adopts a gentle, understated color palette, with light brown tables and wooden wall panels complimenting white brick walls and black furniture. The feel is modern, even minimalist but not pretentious, and the music playing in the background is well chosen but not too curated. The kitchen is hidden, but the bar with pastries and sandwiches is open for display. The staff are smiling and helpful and appear to be proud of their work. The café sprawls into an attractive outdoor space featuring hammocks and deck chairs, which is also a fully functioning urban farm with wicker beds full of vegetable produce.

While offering different examples of Melbourne's café culture, all three of these food enterprises—Grub, Streat, and Louis—are alike in their concern with healthy, good quality food and, as discussed, also share some aesthetic and design qualities and values (postindustrial, DIY), which sit well within the Global Brooklyn framework. However, they differ in the level to which they display their care for sustainable, local, and healthy modes of hospitality. Louis, for example, lets its high-quality, presumably healthy food speak for itself, while the rest of the aesthetic experience is upmarket and minimalist. Grub, which similarly offers a sophisticated menu combining health and cosmopolitan tastes, invites us into a whole garden of plants set outside, while the interior is much more casual and playful. And Streat, as we will see, is in-between but positions the care for the human being and a holistic integrated sustainable food expertise at the center of the experiences they offer.

Whether explicitly or not, we would suggest that many of Melbourne's hip inner-urban cafés also tend to share a certain ethical disposition, leaning toward "do good," socially responsible models of hospitality, even if they are primarily commercial affairs.

Locating Melbourne's Café Scene

Melbourne's thriving café scene has, in recent years, become a central element of the city's "self-understanding and public image" (Cameron 2018). The coffee culture is such that one will regularly hear from locals and tourists that Melbourne has the best coffee in the world. Certainly, while traveling to other café destinations around the world (from Shoreditch to Berlin and Barcelona), we find ourselves regularly informed of Melbourne's status as home to some of the world's best baristas. Even in New York, where we recently researched the rise of Australian cafés in Manhattan and the outer boroughs, we were told that Australians were known as the inventors of "the flat white." While Australian "inspired" and owned cafés (such as the Bluestone Lane cafés that have now spread across the United States—with nineteen locations in New York City alone) are associated with high-quality Italian-style coffee and highly trained baristas, Australians give the Italian coffee culture their own twist.

While perhaps best-known today for its flat white, latte art, and smashed avocado on toast, Melbourne has had a long-standing café culture (Frost et al. 2010; Walters and Broom 2013). While it finds its roots originally in the coffee houses of the nineteenth century, it was the large-scale postwar migration from Italy—Italians are still the second-largest ethnic group in greater Melbourne—that really shaped the city's contemporary obsession with coffee. These migrants largely settled in the poorer northern-inner suburbs including Fitzroy and nearby Carlton, which became known as "Little Italy" and are characterized to this day by a swathe of (now predominantly tourist-oriented) Italian cafés and restaurants. Coffee "empires" like that of Proud Mary, who are coffee roasters as well as café owners, distributing their coffee within Australia and to North America, did not appear *ex nihilo*. Today's hip Melbourne café culture has in part built off and emerged out of Melbourne's Italian café culture, adding flat whites, milk alternatives, decaf, and elaborate latte art to the Italian repertoire of the espresso and cappuccino.

While Melbourne has a distinctive local history and character to its contemporary café, restaurant, and bar scene, which locals see as unique, the city's café culture also shares many of the characteristics associated with what the editors of this collection have termed Global Brooklyn. As we understand it, many authors in this volume are not arguing that café culture (at least in its more bourgeois variations) is becoming the same worldwide or indeed that Brooklyn itself has disseminated its values and taste culture across the world in a top-down fashion. Instead, as the editors are positing in the introduction, Global Brooklyn can be seen as a heuristic or an ideal type characterized by shared values, taste, design, aesthetics, approaches to work and lifestyle, as well as of course shared culinary principles. If we examine inner-urban cafés around the world, from Watthana in Bangkok to London's Shoreditch, using this particular lens (as this collection is attempting), we will find that in many cities there are café cultures that are both distinctively "local," and, at the same time, have a kind of (often uncannily) familiar quality to international travelers. The Global Brooklyn model suggests that the feeling of familiarity we experience in such spaces is related to five shared elements, including *designed* experiences; *sensual* regimes, tied to the local and homegrown as a counter to anonymous corporate food culture; *networked communication*; shared *cultural literacies* across café owners, workers, and consumers alike; and a celebration of *artisanal labor*.

In this chapter, we want to suggest another element in play, that can be similarly found in many such cafés in more or less explicit forms, and which extends the idea suggested in the Global Brooklyn framework that these cafés display varying degrees of connection and resistance to corporate food systems. Alongside the elements discussed at length in other chapters of this volume, we would suggest that another key feature often at work is a kind of "ethics of care." It ranges from a focus on ethical consumption, which expresses itself as care for local producers and for the environment, often tied to a healthy, organic taste aesthetic, to a more deep-seated engagement with social entrepreneurship and alternative models of food provisioning and hospitality. Our position here is an explicitly political one: we want to foreground the conviviality, care, and hospitality found in many of Global Brooklyn's cafés. Our concern is to frame them as potential forms and spaces of experimentation in alternative, ethical social relations: we are interested in the potential that such cafés can have for actual social change or what Gibson-Graham (2006), the pen name shared by feminist economic geographers the late Julie Graham and Katherine Gibson, have termed a "politics of the possible."

An Emergent Postcapitalist Café Culture?

The focus of the establishments that have embraced the Global Brooklyn model has tended to be on the aesthetic and shared design elements of the built (and experiential) environment of cosmopolitan cafés around the world. As Gary Bridge (2007: 39) argues in relation to global gentrification in the built environment, while "The modernist loft would seem to be a million miles away aesthetically from the converted Victorian terrace," the building preferences of a cosmopolitan class "share a common set of aesthetic codes." In spaces like Louis and Grub we do see repeated tropes—aesthetic memes if you like—of certain shared taste and design markers that could be found in cafés anywhere in the world, from Bangkok to Brooklyn, as well as more distinctive Melbourne tropes (nineteenth-century bluestone lanes integrated into café scapes, for instance) that again share a middle-class preference for certain kinds of historicity.

So what kinds of conceptual lenses and approaches might enable us to start to critically engage with and perhaps extend upon the Global Brooklyn model? One useful approach is to draw upon the literature on experimental and alternative economies (Beacham 2018; Gibson-Graham 2014) in what some have termed a postcapitalist environment. Postcapitalism as a term captures a broad spectrum of economics ranging from Peter Drucker's conceptualization of *Post-Capitalist Society* (1994) as centered around knowledge and service workers, to an assortment of participatory and shared models of economics and conceptions of collaborative consumption. In the latter group, for instance, critical marketing and consumer studies scholar Russell Belk (2007) argues that the sharing dynamics underpinning care-based familial exchanges are a robust, ethical, and sustainable alternative to capitalist models based on commodity exchange, noting that such models are flourishing in areas such as online businesses (and, we would add, café culture). As such he is skeptical of arguments that view the sharing and collaborative economy purely in capitalist terms, suggesting that they are marked by broader economies and relations of care.

In fact, cafés can be places with generative potentials for care on many levels. The obvious aspect would be the nourishment of people visiting cafés and consuming food and drink. As Grub's head chef Ravi Presser (quoted on the menu) puts it, "sharing is caring." The craft aspect in cafés tells us about a culture where professionalism operates succinctly and carefully within different modes of hospitality. Melbourne is particularly known for a café culture grounded in the use of exceptionally high-quality ingredients with an emphasis on taking great care in both the preparation and the presentation of food. The ethics of healthy, environmentally sustainable, and socially produced and sourced food is also a key element in this culture of quality with concrete benefits for the health and well-being of customers. How might we understand the role and place of care here and how might this speak to both an alternate politics and economics?

From Care-less-ness to an Ethics of Care

Another food enterprise off Brunswick Street in Fitzroy, which is popular with locals and passing tourists alike, is café Grace. On a street filled with grungy warehouse conversions, artfully graffitied and muralized walls, and slick new townhouses, located near the bohemian Rose Street arts and craft market, Grace stands out from the crowd. Housed in an old standalone cottage that has been converted into an eatery, the café looks and feels like someone's home, with a friendly open student shared household vibe. Grace's customers often sit on the front verandah talking with passing neighbors, patting dogs, and watching the traffic. Out back of the house we find a small "yard," at the rear of which sits a shipping container that has been converted into another small space for the café's customers. The look and feel of the interior of the main part of the café-house is relaxed, with a vintage aesthetic, and often features a mix of dining customers alongside creatives and students working on their laptops while making their soy lattes last as long as possible. While some cafés frown on customers taking up tables in such a fashion, Grace's familiar lounge room projects a welcoming ethos. It is not surprising to see that Grace's motto is "kind food" while their menu uses "kind, quality ingredients." While Grace is a commercial enterprise, we will argue that underpinning a number of the hip eateries in Fitzroy and Collingwood, this prioritizing of kindness marks a certain shift, indicating the potentials for cultures that are not entirely embedded in the logic of capitalism.

The Collingwood-located café Streat translates this ethos of kindness and care into the core of its business model. With the "brand" motto, "Tastes Good, Does Good," Streat is a social enterprise whose focus is to help young, mostly homeless people aged sixteen to twenty-five. Their holistic program enables people to find a job, get a home, feel a sense of belonging, grow as a person, and integrate into the broader society. Streat provides hospitality training, mental health support, and community facilitation. Through integrating people into the café's daily operations across cooking, gardening, customer service, as well as manual labor and mechanical work, Streat provides real jobs and crucially contributes to sustainable livelihoods. More than seventy percent of Streat's income is self-generated through their café, artisanal bakery, and coffee-roasting business. The success and popularity of the café and its model is reflected in the fact that, according to their website (Streat 2020a), in the last year alone their revenue has grown an impressive 31 percent, while the ethical brand has a number of spin-off cafés and café carts around Melbourne, including a hole-in-the-wall café at RMIT, the university at which we both work.

We will get back to both Grace and Streat but first we need to look deeper into concepts of care. In *Design as a Practice of Care*, Laurene Vaughn (2018) defines care as "consisting of performed acts that promote well-being and flourishing of others and ourselves based on knowledge and responsiveness to the one cared for." The practice of care in this sense is grounded in a "connection to the subject of care—either through relationship or empathy"; an "expectation that the actions of care will have an impact or meaning for those being cared for"; and a "capacity to perform the actions of care in a skillful and useful manner" (Vaughan 2018: 12). Caring as such is an act of intimacy and it demands imagination, especially when the act of caring becomes more abstract. "Caring for unknown others is so challenging because the farther we move from the familiar, the less control we have and the more risk of failure there is" (Hamington in Vaughan 2018: 13). This suggests that humility, ambiguity, imagination, and a disregard for the rules are fundamental for a practice of care. Care in this sense becomes a political act with potential radical impacts.

The framework of Global Brooklyn gestures toward a notion of carefulness in the sense of local, healthy food, or even carefully designed interiors. What it does not necessarily capture currently is the political and socially responsive dimensions of "foodie" café culture. We suggest that this involves an extension of the notion of hospitality itself from its more literal meaning within the food services industry to the notion of building a new set of socioeconomic relations around care and hospitality, and in particular caring for the stranger, an ethics that Ghassan Hage (2003) links to a "politics of hope."

The introduction of a sixth element to the Global Brooklyn ideal type proposed in the introduction, which we might understand in terms of an "ethics of care," will therefore enable us to understand a crucial dimension of today's trendy cafés and food cultures. This element takes us beyond a purely "capitalocentric," brand-based reading of some of the collective values we see emerging in Global Brooklyn establishments. That is, the forms of food citizenship and modes of care-based sociality performed in these cafés cannot be assumed to purely fit into a top-down model of advanced capitalism but suggests that a range of other logics could be at play. For instance, American celebrity farmer-activist Wendell Berry (2009) argues that today "eating is an agricultural act," highlighting "the fact that what we put on our plates today is linked to a wider politics of animal welfare, sustainability, food security, the plight of producers in developing nations in the face of globalizing agri-business as well as questions of food safety, health, and risk" (Lewis 2020: 5). Similarly, we argue that café culture and eating in social spaces has become a complex reflexive set of social practices, linked to questions of where food comes from, who prepares it and how, and what the act of hospitality itself might say about our broader social relations.

The ethics of food businesses, the broader practices of social eating and conviviality, as well as the associated work practices that they promote are particularly important to look at closely because they reflect socioeconomic relations, conventionally tied to modes of capitalism but increasingly gesturing toward other forms of "governance," systems of value and rules for living (Vodeb 2017). If certain kinds of fat- and sugar-saturated convenience food and eating are tied to capitalist models of pleasure, which Vodeb (2017: 484) has argued are embedded in a "drug-like" model of habituated use, in the bourgeois taste cultures of Global Brooklyn we also see other models of pleasure at play, tied to forms of mindfulness, reflexivity, and care, that is, paying attention to the impact of eating and food itself. The contradictions at play in these cultures suggest the existence of new micro levels of productive tensions, which open up new possibilities for alternative food practices.

The French philosopher Bernard Stiegler (2012) argues that while (and because) capitalism is in crisis it has also attempted to develop a new approach to governance or population management that sees the state moving from Foucault's notion of "biopower" to what Stiegler himself refers to as "psychopower." Of particular interest to us here is his argument that psychopower is based on a state of *care-less-ness* through a systematic destruction of attention (Stiegler 2012). The purpose of this power is to both produce and control consumers, destroying and dominating attention through what Stiegler calls psychotechnologies. Vodeb links this process of destruction of attention more concretely to food, drugs, and/or technology (Vodeb 2017). But food as we see it in the context of new and emergent café cultures may also be linked with carefulness and hospitality that in the case of enterprises like Streat can have a transformative social impact. As Stiegler points out,

> [o]n the side of consumption, the capitalist mode of life has become at the end of the 20th century an addictive process less and less capable of finding sustainable satisfactions—this has induced great discontent in the civilisation of consumption, which has replaced culture, that is, care, if we accept that culture precedes cults of all types, that is, attachments to objects whose ensemble constitutes a system of care. (2012: 111–12)

New Labor: Work and Creative Labor as Care for Others

One of the elements highlighted by the Global Brooklyn ideal type proposed in the introduction is the way in which once devalued (and increasingly industrialized and alienated) forms of manual labor are being revalued as artisanal, craft work. In cafés like Industry Beans, 100 meters from Grace, we find ourselves in a former industrial warehouse space that combines a slick postindustrial aesthetic and a sophisticated high-quality food menu with an active coffee roastery and brew house. Indeed, an essential part of the café experience is being able to view aspects of this process, with the back area of the café full of large sacks of coffee beans. As we see from the following quote from the café's website, the industry of Industry Beans is "redefined" as a kind of intelligent and informed labor, and as a "vision" and a "movement."

> Over the last two decades, a wave of coffee roasters across the globe, known as the Third Wave, redefined ideas about sourcing, roasting and brewing coffee. [. . .] Inspired by this movement, brothers Steven and Trevor Simmons founded Industry Beans in Melbourne in 2010. Their vision was to roast high-quality specialty coffee in a carefully curated environment of transparency and accessibility, to offer the ultimate coffee experience to their community. (Industry Beans 2020)

As we have suggested above, this focus on labor and quality of craft is more than about branding and business enterprise. The playful and almost ironic use of the term "industry" in Industry Beans marks a self-aware distanciation from massified, factory-oriented production methods and the care-less-ness that marks the realm of generic coffee manufacturing. Revealing to customers the protracted, thoughtful, and meticulous methods by which the coffee they drink is made, is also about performing an act of care.

This association between acts of thoughtful, artisanal labor and broader questions of care reveals, as demonstrated by sociologist Richard Sennett's work on materiality, craft, and culture building, a very basic human impulse toward countering processes of alienation (Sennett 2008). This humanizing and vitalizing impulse is also observed in an Instagram post by Streat. Next to a highly appetizing photo, which shows tender almond-covered pastries that have been skillfully sugar coated by an expert pastry chef, the text reads: "You've made it to Friday, now celebrate with one of our tasty pastries. A lot of care and attention goes into making these show-stopping pastries at our Cromwell St bakery; it's a four-day process from start to finish. Get yours at one of our cafés" (Streat 2020b). There is an interesting dialectic here between the chef's labor and the pastry as a reward for the customer themselves making it through the working week. That tension between work and pleasure, or reward, is also evident in the café experience itself where being an ideal Streat customer—with knowledge of the complex ethical menu and of the social enterprise model of the café—requires an extensive commitment of time and effort. Labor, pleasure, and time, however, are differently inflected here through social relations of care. In contrast to fast food and convenience cooking and consumption, slowness, care, and attention itself are embedded in the very craft of artisanal pastry making, which in turn demands the undivided attention of the café's guests via its "show-stopping" quality. Can such an act of cooking, food preparation, presentation, and hospitality channel carefulness to the guests, patrons, customers of Streat, and perhaps the wider society via a different model of social relations to Stiegler's capitalist realm of distracted attention and care-less-ness?

In a 2016 essay for the *New Left Review*, philosopher and feminist theorist Nancy Fraser, like Stiegler (2012), argues that financialized capitalist societies experience what she calls a "crisis of care." Here she is particularly concerned with capitalism's historical reliance on social reproductive labor performed in the home. Unpaid (and often historically undervalued), this labor nevertheless forms the bedrock of society and the economy (Fraser 2016: 108–12). With women increasingly working full time, however, domestic labor has become a scarce resource, a transition that has seen a range of developments occur from outsourcing child care to bringing wage labor increasingly into the home. On the other hand, we are also seeing a kind of extension of feminine modes of care into spaces once seen as primarily sites of business and paid service, and in particular into café spaces. Certainly, with global pressure on space in urban areas, and with the rise of vestigial kitchens in apartment dwellings, cafés become quite literally an extended home space away from home, as reflected in cozy, domesticated spaces like Grub (which features old couches and a lounge-like chill space), and Grace's literal use of a house as a site of both enterprise and kind hospitality. The work performed in these spaces is still of course labor—and anyone who has worked in hospitality knows how hard that work can be—but it is work that both revalues care *as work* and as wage labor. There is a large body of scholarship, from Arlie Hochschild to Eva Illouz that reads the re-valuing within neoliberal societies of this kind of gendered care work as an extension of the logics of capitalism into our embodied and emotional lives. There is much to be said for this argument but it also ironically represents a tendency to read all forms of sociality in capitalocentric terms. Our argument here is that in the experimental spaces of new forms of café culture, we are also seeing care labor as being premised on more than just the extraction of value.

Another form of care as work that we can observe is also manifesting itself in the time and effort invested by patrons, visitors, and consumers of such ethical cafés, something we can see as an extension of the broader ethicalization and politicization of consumerism that has occurred in many capitalist democracies around the world (Barnett et al. 2011; Lewis and Potter 2011; Littler 2009; Micheletti 2003; Soper, Martin, and Thomas 2009). The information provided on the reverse side of the A3-sized and carefully designed menu at Streat is complex, demanding, and takes substantial time to read, especially in a café setting. There is an obvious expectation by such establishments that the visitor already is significantly educated on matters of food democracy and of course has a sophisticated palate. The menu itself offers an elaborate range of cosmopolitan dishes from Macanese spicy Minchi to Mochi donuts with miso caramel, mandarin curd, yuzu sesame and freeze-dried mandarin, and the mind-bending Bloody Madam cocktail with Fair quinoa vodka. The topics presented in the menu range from a presentation of Streat's business model to their specific achievements like full renewable energy use, and a work philosophy that shows a complex set of relations and practices touching all levels of the food system. Among other impressive achievements we can read, from 2018 to 2019: "Magic our Labrador/Kelpie therapy dog continued to work her magic and provided 1,920 Therapy Dog hours to our people and staff."

In Conclusion: Rethinking the Economy through Café Culture

The statistics and infographics on Streat's menu reflexively gesture on the one hand to the kind of Key Performance Indicators (KPIs) we are now familiar with in neoliberal cultures, while on the other they reference a very different kind of exchange and value, one embedded in care, well-being, and the strengthening of social relations. Gibson-Graham have written extensively about the emergence of alternative and community-based economies. Their research—indicating that people around the world are experimenting with other ways of doing economics and work in everyday life—is central for understanding the experiments with care and kindness we are seeing emerging in cafés in Melbourne and in international Global Brooklyn-style establishments more broadly. In highlighting the ethical dimensions of cosmopolitan café culture, we see ourselves as contributing to an alternative reading of the "life politics" underpinning this turn, linking the focus on creativity, community building, and an "art and aesthetics of everyday living" to modes of ethical citizenship (Lewis 2012: 2).

As Gibson-Graham have noted (2014), a key problem with economic scholarship on social change is that it tends to be premised on "strong capitalocentric theories of economic reorganization," which means that the ethical turn in once highly commercialized environments like hospitality tends to be too easily written off as purely being about branding and product differentiation. Fostering a "politics of the possible" (Gibson-Graham) then calls for moving from what Boaventura de Sousa Santos (2001) terms a "sociology of absences" to documenting "ecologies of difference" (Gibson-Graham 2014). As theorist Franco Berardi argues, we need to wrest the social imaginary away from neoliberal capitalism and develop a new "sensuous" language of "meaning and desire" that cannot just be reduced to monetary exchange (Berardi 2012). Part of the work that we are doing in this chapter then is to make visible and to legitimate—through focusing on the nuances and contradictions as much as new possibilities—other ways of framing the developments in café culture that this collection is debating within the framework of Global Brooklyn.

This involves a combination of experiential ethnographies of the local and an open approach to theories of sociality, economics, and change, or "weak theory," as Gibson-Graham term it.

> It is ethnographic thick description and weak theory that have helped us to imagine the becoming of community economies motivated by concerns for surviving together well and equitably; distributing surplus to enrich social and environmental health; encountering others in ways that support their well-being as well as ours; maintaining, replenishing, and growing our natural and cultural commons; investing our wealth so that future generations can live well; and consuming sustainably. (Gibson-Graham 2014: S152)

Thus, we have looked at the dynamics of change as an open empirical question rather than a structural imperative, and through that process we are attempting to articulate a different imaginary. Of course, the cafés we have discussed above do not all fit into the more utopian moment of not-for-profit equitably imagined in the abovementioned quote. These cafés range from fairly standard capitalist enterprises such as Louis and establishments with a social conscience to alternative forms of paid labor and socialized market transactions as evidenced in Streat. Industry Beans, for instance, is donating ten percent from all coffee bean sales to the Australian Red Cross, a humanitarian organization actively involved in supporting people affected by the 2019/2020 Australian bushfires. Similarly, the bourgeois taste cultures of Global Brooklyn more broadly still tend to be embedded within capitalist modes of business even as they embrace elements of ethical consumption and purchasing and in some cases ethicalized and socialized labor practices. However, by paying attention to the noneconomic social practices increasingly central to these hip cafés (community building, conviviality, creativity, dynamics of sharing, and gift exchange), we can also detect a strong manifestation of a maturing ethics of care. We predict that this dynamic will develop further in the future and contribute significantly to much needed ethical, environmental, and social change.

Acknowledgments

We would like to thank Dr. Nicholas Hill for his excellent research and editorial support.

References

Barnett, C., Cloke, P., Clarke, N., and Malpass, A. (2011), *Globalizing Responsibility: The Political Rationalities of Ethical Consumption*. Malden: Wiley-Blackwell.

Beacham, J. (2018), "Organising food differently: Towards a more-than-human ethics of care for the Anthropocene," *Organization*, 25 (4): 533–49.

Belk, R. (2007), "Why notshare rather than own?" *The ANNALS of the American Academy of Political and Social Science*, 611 (1): 126–40.

Berardi, F. (2012), *The Uprising: On Poetry and Finance*. Los Angeles: Semiotext(e).

Berry, W. (2009), "Wendell Berry: The pleasures of eating," *Centre for Ecoliteracy*. Available online: https://www.ecoliteracy.org/article/wendell-berry-pleasures-eating (accessed January 28, 2018).

Bridge, G. (2007), "A global gentrifierclass?" *Environment and Planning A: Economy and Space*, 39 (1): 32–46.

Cameron, A. (2018), *Affected Labour in a Café Culture: The Atmospheres and Economics of "Hip" Melbourne*. London: Routledge.

Drucker, P. (1994), *Post-Capitalist Society*. New York: Harper-Collins.

Fraser, N. (2016), "Contradictions of capital and care," *New Left Review*, 100: 99–117.

Frost, W., Laing, J., Wheeler, F., and Reeves, K. (2010), "Coffee culture, heritage and destination image: Melbourne and the Italian model." In L. Joliffe (ed.), *Coffee Culture, Destinations and Tourism*, 99–110. Bristol: Channel View Publications.

Gibson-Graham, J. K. (2006), *A Postcapitalist Politics*. Minneapolis: University of Minnesota Press.

Gibson-Graham, J. K. (2014), "Rethinking the economy with thick description and weak theory," *Current Anthropology*, 55 (9): S147–S153.

Hage, G. (2003), *Against Paranoid Nationalism: Searching for Hope in a Shrinking Society*. Sydney: Pluto Press.

Industry Beans. (2020), "Our story." Available online: https://industrybeans.com/pages/our-story (accessed February 24, 2020).

Lewis, T., and Potter, E., eds. (2011), *Ethical Consumption: A Critical Introduction*. London: Routledge.

Lewis, T. (2012), "'There grows the neighbourhood': Green citizenship, creativity and life politics on eco-TV," *International Journal of Cultural Studies*, 15 (3): 315–26.

Lewis, T. (2015), "'One city block at a time': Researching and cultivating green transformations," *International Journal of Cultural Studies*, 18 (3): 347–63.

Lewis, T. (2020), *Digital Food: From Paddock to Platform*. London: Bloomsbury.

Littler, J. (2009), *Radical Consumption: Shopping for Change in Contemporary Culture*. Berkshire: Open University Press.

Micheletti, M. (2003), *Political Virtue and Shopping: Individuals, Consumerism, and Collective Action*. New York: Palgrave Macmillan.

Ritzer, G. (2009), *The McDonaldization of Society*. Los Angeles: Pine Forge Press.

Santos, B. (2001), "Nuestra America: Reinventing a subaltern paradigm of recognition and redistribution," *Theory, Culture & Society*, 18 (2–3): 185–217.

Sennett, R. (2008), *The Craftsman*. New Haven: Yale University Press.

Soper, K., Martin, R., and Thomas, L. (2009), *The Politics and Pleasures of Consuming Differently*. Basingstoke: Palgrave Macmillan.

Stiegler, B. (2012), "Care." In T. Cohen (ed.), *Telemorphosis: Theory in the Era of Climate Change*, 104–20. Michigan: Open Humanities Press.

Streat. (2020a), "2019 Annual Report." Available online: https://www.streat.com.au/site s/default/files/191202-str_2019_annual_report.pdf (accessed February 24, 2020).

Streat. (2020b), "Instagram post." Available online: https://www.instagram.com/p/ B43DQLRhjav/ (accessed February 15, 2020).

Vaughan, L. (2018), *Designing Cultures of Care*. London: Bloomsbury Visual Arts.

Vodeb, O. (2017), "Conflict kitchen." In O. Vodeb (ed.), *Food Democracy: Critical Lessons in Food, Communication, Design, Art and Theoretical Practice*, 484–513. Bristol: Intellect Books UK.

Walters, P., and Broom, A. (2013), "The city, the café, and the public realm in Australia." In A. Tjora and G. Scambler (eds.), *Café Society*, 185–205. New York: Palgrave Macmillan.

Dispatch

Global Zen and the Art of Local Coffee: Japanese Cafés in the Age of Global Brooklyn

Helena Grinshpun

A simple white facade, a sliding glass door with the menu neatly pasted to it, few chairs outside; the narrow inner space greets the visitor with today's drip coffee recommendation scribbled on a chalkboard. The small seating area is comprised of the counter and a couple of tables cramped along the wall; behind the counter is the barista, surrounded by large jars with coffee beans, smaller jars with homemade cookies, yet smaller jars with spoons and straws, and mason jars for serving iced coffee. Shelves are filled with a colorful mismatch of objects—vinyl records, old posters and magazines, figurines, old cameras, and other memorabilia. A few table lamps create a sense of quiet coziness even on a busy afternoon, when the narrow streets of trendy Shimo-Kitazawa, where the café is located, are bustling with people. The café is named after the owner's favorite band's album "Bookends," plays Western classic pop and rock music, and sells manually brewed specialty coffee at surprisingly reasonable prices. Although its history does not go back any further than several years, the partnership with a long-running local roaster grants its coffee the quality of tradition so highly valued in Japan. Moreover, the owner sees himself continuing the legacy of Japan's early coffee shops that introduced culture through music. Born in the late 1950s, he remembers how the cafés were places for people to get together and talk. That is why Bookends Coffee Service is open until late and frequently hosts live music events despite the lack of space.

Let us shift our gaze away from fashionable Tokyo, where good coffee is hardly a rarity, to the sacred island of Miyajima, nested in the picturesque waters of Hiroshima Bay. Covered with lush vegetation and a home to several important temples and shrines, Miyajima is anything but urban, trendy, or hip. Yet recently, its small retail area, filled mainly with souvenir shops and local eateries, surprises the visitor with stylish coffee shops serving locally roasted specialty coffee. One of Miyajima's coffee pioneers, Café Sarasvati, hides its tasteful premises behind a dark wooden facade. Heavy doors lead to a dimly lit spacious room with dark-painted walls, concrete floor, plain timber tables, and naked light bulbs hanging from the ceiling. A small wooden stand displays packages with beans and blends sold at the café; a rustic copper scale adds an arty touch. A coffee-roasting machine, although silent at the moment, implies on-site roasting. Manual coffee-making equipment is arranged on the shelf behind the barista; the wooden counter is almost empty, inviting those coming for a take-out to lean over and watch the baristas prepare the order. Wooden stairs lead to the upper floor, where a light buzz of conversation engulfs those dining on a nicely crafted cake or a lunch set. Every now and then a non-Japanese tourist peeks into the café, spots the espresso machine, and leaves a few minutes later, sipping happily from a recycled paper cup filled with aromatic espresso. Only a few years ago, espresso would be unthinkable in a place like Miyajima, says a young male barista who commutes to work from the nearby Hiroshima city. Now, new cafés are opening, infusing the island's *shibui* (tastefully old-fashioned) atmosphere with trendy spots offering great coffee in a fashionable setting.

Let our gaze now rest on yet a different site—Café DOnG by Sfera, situated in the historic district of Gion, in the old capital of Kyoto. Its minimalistic and sophisticated interior skillfully combines traditional Japanese motifs with contemporary product design—natural materials, clean lines, clear colors, potted plants scattered in a manner that excludes anything accidental. Along with regular tables and chairs, an elevated floor with low tables and cushions allows customers to enjoy a traditional Japanese seating style. The menu is seasonal and includes freshly ground coffees along with green teas, spirits, locally produced traditional sweets, organic food, and vegan dishes. As suggested by its rather unusual name, the café is designed by Sfera, a Kyoto-based brand focusing on interior accessories and aspiring, according to its website, to create a "modern fusion of time-honored materials, traditional craftsmanship and contemporary aesthetics."[1] Sfera shop and gallery occupy the floors above the café in the purpose-built Sfera building, renovated recently by a Swedish architect.

The three cafés are worlds apart but belong to Japan's changing, increasingly diverse, and sophisticated coffee scene. While shifting, however, this scene is far from being new. Contrary to its image as a tea-drinking country, Japan has known coffee for more than 150 years; coffee shops (*kissaten*) have been for decades an integral part of Japan's landscape and lifestyle. Japanese coffee culture developed against the backdrop of the country's modernization and Westernization. Coffee was first brought to Japan by Dutch traders in the seventeenth century, but did not become a familiar commodity until the end of the nineteenth century, when, after more than 200 years of seclusion, Japan was forcibly opened to Western trade. Along with other foreign items and trends, from meat to educational reform, coffee became associated with civilization and modernity. The first coffee shops on Japanese soil opened in the 1880s and were modeled on the European café. Besides coffee, they offered Western architecture and décor, Western furniture, and Western food, served by waitresses dressed in Western clothing. Most carried European-sounding names, such as Café Printemps, Café Lion, Café Paulista, and Ginza Palace, among others.

In the 1920s and the 1930s, coffee shops already became a visible segment of urban life. Although most of them had to close toward the 1940s, the postwar era witnessed a renewed infatuation with coffee. Japanese society had undergone numerous transformations in which the United States served as a frame of reference. One manifestation of this influence was the popularity of "jazz cafés" that provided the means of coming in contact with the world by playing American jazz music. By the 1980s coffee shops were booming, offering different styles of coffee, various gimmicks, atmospheres, and styles of décor. The 1990s brought new trends such as the proliferation of global coffee chains. The 2000s were marked by a growing exposure to global media, and, as a result, global consumer images and fads, which determined a further diversification of coffee-related practices and tastes.

A century after the first café opened in modernizing Japan, the local coffee scene came to be characterized by several key elements. First, despite its acquired banality, coffee still possesses a certain foreign flavor. There is a vast use of foreign terms in the coffee business, written in either English or *katakana*, an alphabet used for loanwords. The décor of the café tends to incorporate images and decorative patterns referring to non-Japanese, mainly Western cultural contexts. The outfit of the barista often includes a bow tie and other items of formal European attire; porcelain cups and saucers along with silver teaspoons, milk jugs, and sugar bowls constitute a setting rather atypical of a traditional table. Over the decades, the *kissaten* developed a basic food menu perceived to be appropriate for coffee, most of which is associated with Western cuisine, such as spaghetti and toast.

The soundscape of the coffee spaces is usually comprised of non-Japanese classic genres, such as jazz, blues, and classical music. Jazz and classical music in particular have come to be firmly associated with coffee shops, often defining a formal framework for public coffee-drinking. Jazz cafés and *meikyoku* (classical music) cafés, although less common today than a few decades ago, still offer their customers a nostalgic get-away into the world of analog sound and strict etiquette that "make time move slowly" (Plourde 2019: 43). While less total in their sonic environment, many coffee shops play classical background music; some, such as Café Verdi in Kyoto, host live music concerts.

Another attribute of Japan's coffee culture is its long-standing emphasis on manual preparation of coffee, and, as a result, on the spectacle of coffee making. This emphasis manifests itself as an elaborate aesthetic regime centered on the performance of the coffee "master." Merry White (2012) describes how the Japanese coffee masters exercise authority and affectionate obligation to customers through acting as both performers and mentors. Many of the cafés are characterized by intense *kodawari* (meticulous attention to detail) in everything that has to do with the coffee experience—specialization in particular blends, the choice of serving utensils, the use of brewing and roasting equipment, and the performance of a complicated set of tasks by the coffee master. The performance is carried in front of the customer, turning coffee preparation into a theatrical act, one cup at a time.

The preference for drip coffee and manual brewing, along with a preoccupation with mastery, fostered the development of specialized coffee equipment—hand drippers, siphon coffee makers, pouring kettles, and other paraphernalia. In order to appreciate the coffee crafted through the alliance of masterly tools and masterly skill, a certain level of consumer expertise is necessary; it is often mediated by the master, who assumes the role of a coffee mentor.

The coffee experience is designed by the coffee shop master according to his (rarely, her) individual vision. The cafés therefore embody individual stories and fantasies, making each site unique despite the common denominators. This, perhaps, determined the successful indigenization of coffee in Japan—as meaningful local sites, cafés have become a part of collective memory and identity.

The Japanese coffee scene is constantly changing. Old-style *kissaten* run by an owner-master with a bow tie still dot the narrow streets of Japanese cities. However, new forms are emerging—indie cafés with organic slow food menus and trendy espresso bars, thematic cafés with selfie corners for Instagram users, and library cafés for book lovers. The makeup of these spaces frequently follows the recognizable lines of the aesthetic regime referred to in this volume as Global Brooklyn—exposed pipes and bare concrete, mismatched furniture, well-designed plainness, preference for fresh, seasonal and health- and environment-friendly ingredients and, of course, artisanal coffee, all making one wonder whether she is in Osaka or Oslo, Brooklyn or Tel Aviv. Vegan dishes are as new to Japan's coffee shop scene as are recycled paper cups; espresso-based beverages and latte art have started making their way into the local coffee cups only recently. The same is true about the use of computers—for decades, *kissaten* provided a space for a coffee, cigarette, and newspaper, creating sites for recreation rather than productivity. Today, laptops—especially MacBooks—have become an integral part of the café cool, with its affinity between coffee and gig labor framed by the recent term "coffice" (Droumeva 2017).

The novelties, however, appear to embrace much continuity. While the overwhelming preference for drip coffee is slowly giving way to infatuation with espresso, it does not undermine the centrality of a spectacle. The barista's performance is always visible, put on show to be enjoyed and appreciated. As articulated by one of my café-goer interviewees, in Japan your coffee always has a face.

The West, as Japan's long-term cultural "other," still provides a frame of reference, even if less apparent. Leather chairs and pin-up art, iconic photographs of Bob Marley and the Empire State Building, checkered blankets, blackboard menus displayed in English or French or Italian, as well as foreign-sounding names of the café, constitute a pastiche that has come to be associated with a generic café setting. Lately, this pastiche often incorporates references to local traditions: latte served in Japanese ceramic cups with charcoal spoons, menus inspired by local flavors and ingredients, interior design based on minimalistic Zen (or what has come to be globally perceived as Zen) aesthetics. The website of the abovementioned Café Verdi opens with a YouTube video featuring a *maiko* (*geisha* apprentice) sipping coffee to the sounds of classical cello and narrating a basic lesson in drip coffee preparation. Café Sarasvati's homepage features a description of Miyajima's unique environment grounded in a harmonious relationship between religion and nature, which determines the superior quality of local coffee.

The emphasis on performance, manual equipment, and acquired taste often links Japanese coffee with another aesthetic regime—the tea ceremony. Having evolved from an elitist practice of the premodern era into a marker of Japan's cultural heritage, tea ceremony has become a world-familiar icon of Japanese aesthetics. The cultural tie between coffee and tea, increasingly exploited by the media, informs the constant quest for authenticity by presenting coffee as anchored in a long tradition of refined craft.

Originally imported from the West and indigenized as a foreign novelty, Japanese coffee culture is now making its way back to the West, dictating trends of artisanship to the global coffee world. In recent years, Japanese-style cafés have been opening in big cities of North America and Europe, branded as a distinct gustatory and aesthetic experience fitting into the "third wave" agenda. Several Japanese companies, such as Hario and Kalita, lead the market in the supply of low-tech, high-fidelity equipment for by-the-cup coffee brewing. The *kodawari* inherent in Japan's attitude to craftsmanship is said to have helped bring about a coffee renaissance by shifting the focus to a slow, manual, and artisanal—and therefore authentic—way of brewing coffee.

Ogawa Coffee, one of the oldest coffee house chains in Kyoto, has opened its first overseas store in downtown Boston. Making use not only of Japanese coffee brewing techniques and equipment but also of distinctively Japanese design motifs (such as sakura blossom latte art), it describes itself as an "urban oasis, a cute sanctuary . . . of exquisite quality and precision artistry."[2] Hi-Collar (Haikara) coffee shop in New York's East Village promotes itself as a "Western-inspired Japanese café"[3] aspiring to introduce the slower, more meticulous, and leisurely *kissaten* style of coffee shop to the hurried world of coffee in the city. With its dark wooden panels and rustic design, the place recreates the atmosphere of Japan's old-style coffee shops. The slogan of Hi-Collar, "Flirting with the West," acknowledges the peculiar cultural trajectory of Japanese coffee. Similar to the manner Japan's early cafés adopted foreign-sounding names and recreated Western-looking setting, today Japanese cafés abroad hire Japanese employees and make use of Japanese names, dishes, and designs.

The global authenticity of Japanese coffee is informed by its record of a civilizing commodity re-civilized through an association with old local traditions. Now, this cultured commodity is being incorporated into a transnational framework of newly civilized postindustrial consumption. While making a point of blending into the cosmopolitan setting of the postmodern city, Japanese cafés continue to highlight their Japanese flavor, granting their product both cultural specificity and global fluidity.

Notes

1 Sfera official website, available at http://www.ricordi-sfera.com/en/about/

2 Ogawa Coffee USA official website, available at http://www.ogawacoffeeusa.com/

3 Hi-Collar official website, available at https://www.hi-collar.com/.

References

Droumeva, M. (2017), "The coffee-office: Urban soundscapes for creative productivity,"
    BC Studies, 195: 119–27.

Plourde, L. (2019), *Tokyo Listening: Sound and Sense in a Contemporary City*.
    Middletown: Wesleyan University Press.

White, M. (2012), *Coffee Life in Japan*. Berkeley: University of California Press.

3

Copenhagen

Porridge Bars, Nordic Craft Beer, and Hipster Families in the Welfare State

Jonatan Leer

Along with the rest of the Nordic region, Copenhagen had suffered from a poor culinary reputation until the development and success of the New Nordic Cuisine in the early 2000s (Leer 2016). This cuisine was conceptualized in a 2004 manifesto (Skårup 2013). The ambition was to create a cuisine based exclusively on ingredients from the Nordic region. The food should be sustainable, seasonal, lighter, and more vegetable-based than the traditional "meat, potatoes, and gravy" ideal dominating the Nordic region in the twentieth century (Jensen 2014). Although these novel ideas may sound like the French "nouvelle cuisine" from the 1970s (Jönsson 2013), the goal was to challenge the dominance of the French and Mediterranean cuisine by focusing exclusively on Nordic produce.

Restaurant NOMA was the flagship of the New Nordic Cuisine movement and a very successful one. It was voted the best restaurant in the world in 2010, 2011, 2012, and 2014 by the Restaurant Magazine's World's 50 Best list. This sparked a culinary revolution in Copenhagen and the Nordic region in general. The number of Michelin-starred establishments in Denmark went from eight in 2003 to thirty-four in 2019, most of them close to the capital. This development has also had a positive spin-off effect on the mid-level food scene with an impressive amount of restaurants and, lately, with less traditional concepts like Reffen, the new street food and craft market in a former shipyard which was noted in the *Lonely Planet* guide's description of Copenhagen as one of the regenerated urban areas with "indie places" (Lonely Planet 2019). Reffen and similar areas draw heavily on the Global Brooklyn aesthetics and mythology.

In this chapter, I will analyze these Nordic versions of Global Brooklyn in Copenhagen based on field work conducted in 2018–19. I will focus on two newly gentrified areas where this trend is particularly strong, namely the area around Jægersborggade in the Nørrebro neighborhood and the abovementioned Reffen food market.[1] I focus on these two areas as they represent two different kinds of trendy food culture with a rather distinct aesthetic and atmosphere. Thus, this comparison allows for nuanced analyses and discussions of the variations in the Copenhagen Global Brooklyn food scene. I will highlight how many ideal-type ideas discussed in this volume's introduction recur in Copenhagen, but also how the Global Brooklyn spots in the city—contrary to most other versions of this aesthetic regime—are adaptable to middle-class family life. These families incarnate the fusion of national welfare policies and cosmopolitan urban aesthetics and lifestyle. On the basis of these analyses, I will discuss the intricate relationship between the fine-dining scene and the Global Brooklyn establishments and propose a distinction between a chic and a rough version of Global Brooklyn.

As we shall see in this chapter, a complicated relationship exists between the New Nordic movement and the Global Brooklyn aesthetics in Copenhagen. There are many similarities between the core values of the former and the ideal type identified for the latter, among which the nostalgic revitalization of old food crafts, the focus on authenticity and wholesome production, the disapproval of convention and industrial food production, as well as skepticism of homogenized taste in the contemporary food system. At the same time, there are distinct differences. Firstly, a certain territorial "chauvinism" (Counihan 2004) seems to pervade the New Nordic movement, as it refuses to use products from outside the Nordic region (Andreassen 2014; Neuman and Leer 2018). This contrasts with the cosmopolitanism of the Global Brooklyn sensibility. Secondly, in terms of sensorium, the New Nordic Cuisine aims to highlight purity and freshness above everything else (Haraldsdóttir and Gunnarsdóttir 2014). The Global Brooklyn palate seems more oriented toward challenging and complex flavors. Furthermore, many of the core elements in Global Brooklyn, such as the obsession with coffee, is absent in New Nordic Cuisine due to geographical restrictions. Thirdly, the type of nostalgia differs. New Nordic Cuisine is built around a desire to restore the bond with pure Nordic nature, generating a site-specific and nature-centered nostalgia. The nostalgia of Global Brooklyn relates to restoring craft and creativity to the postindustrial cityscape. It is a generic and urban-centered nostalgia. In terms of aesthetics, this also means that New Nordic Cuisine restaurants (both fine dining and mid-level) tend to use natural features in their interiors, often impeccably curated and striving toward the reflection of nature. Global Brooklyn establishments, on the other hand, tend to be unpolished, unfinished, and more heterogeneous and, very importantly, they favor the urban materials of iron and cement. Wood seems to be a material acceptable to both regimes.

Jægersborggade

The Nørrebro neighborhood in Copenhagen is close to the inner city. It used to be a working-class community, and since the 1970s, many immigrants have moved there. The area was also a hotbed of leftist and anarchist activism. Since the 1990s, Nørrebro has increasingly been renovated and subsequently gentrified.

The Street of the Underdog

Jægersborggade is situated in what used to belong to the sketchier parts of Nørrebro, as emphasized in a 1993-reportage called "The Street of the Underdogs": "It used to be a working-class street. Then it became a street for the biker gangs, and today it is populated by the unemployed, students, immigrants and the elderly" (Lisberg and Thorsen 1993; my translation). The reportage continues with colorful descriptions of the street and its many outcasts, notably groups of alcoholics on benches drinking in public and the apparent drug-dealing.

The same year, nonetheless, many of the apartments in the street changed status from social housing and rental apartment to "andel" apartments, a particular Danish system of collective cooperatives, which requires the inhabitants to have a stable income. This development, along with a publicly funded street renovation, changed the demography of the street (Krogsgaard and Madsen 2010). Many of the least affluent people could no longer afford to live there. Subsequently, the street attracted a lot of "young professionals" (mostly well-educated couples) aged between thirty and thirty-four, who constituted 22 percent of the population in 2005 compared to 11 percent in the early 1990s (Krogsgaard and Madsen 2010: 768). New trendy shops and concept stores opened in the street to satisfy the demands of that new adventurous middle-class population. The rent for businesses was still relatively cheap by Copenhagen standards.

Relæ and Puglisi

There were still a few drug dealers hanging around when the chef Christian Puglisi opened his restaurant Relæ in 2010. Puglisi had been sous chef at NOMA for a couple of years under the celebrated head chef René Redzepi of NOMA. The ambition with Relæ was to break out of the New Nordic Cuisine dogmas (and the shadow of NOMA) and to propose a more globally oriented and casual cuisine liberated from the formality of the Michelin-style restaurants (Leer 2016).

In his book *Relæ: A Book of Ideas*, Puglisi describes Jægersborggade in 2010: "The rugged atmosphere of the street had attracted a few creative entrepreneurs to begin with, such as the soon-to-be-famous Coffee Collective and Elsgaard, maker of handmade shoes" (Puglisi 2014: 22). When Puglisi followed, "many considered us lunatics to think of opening a restaurant there" (Puglisi 2014: 22). At the opening in 2010, the restaurant and Jægersborggade found themselves at the heart of a drug war. But this did not stop the flux of new (often food-centered) businesses from setting up shop in the street. Then came the media attention. Since the early 2010s, the street has been one of the trendiest in Copenhagen and available spots have been in demand (Puglisi 2014: 23). A central part of Puglisi's narrative (which returns in the storytelling of other food entrepreneurs, as we shall see) is the bold move to open a restaurant in this "dangerous" street on the edge of society. This evidently adds to his image of a truly adventurous and alternative entrepreneur. It distances him ideologically and physically from the rest of the culinary elite.

This distance and edginess also serve to affirm Puglisi as one of the chefs who wanted to establish a restaurant serving good food, but without the snobbishness of the world of fine dining. Puglisi wanted to put all his energy into tasty, simple food at an affordable price by cutting down on service and waiting staff—and hence also the formality of fine dining (Leer 2016: 8). Soon after opening Relæ, Puglisi opened the bistro Manfreds across the street with an even more casual ambience.

Puglisi proposed an alternative to NOMA and the aspiring fine-dining scene with a Nordic focus in Copenhagen, both in terms of food, atmosphere, and service design (Puglisi 2013). Puglisi is of Norwegian and Italian descent and considers himself "a child of a globalized world." He wanted his food to reflect this cosmopolitan identity and not be restricted by a narrow Nordic ideology of locavorism (Puglisi 2013: 33). Also, the restaurant was one of the first to embrace natural wines (often produced in Denmark) as an alternative to traditional fine wines, challenging consumers with their untamed, unusual flavors. Again, this is a mark of edginess and stepping back from mainstream dining on the Copenhagen restaurant scene at the time. Today, Manfreds has a great window display featuring a colorful collection of natural wine bottles.

Similarly, the nearby wine bar and wine shop Terroiristen on Jægersborggade offers 100 percent natural wines. The natural wine trend has spread rapidly across Copenhagen and on Jægersborggade, so much so that it may now be difficult to find conventional wines.

Like the many underdogs who used to live in Jægersborggade, Puglisi himself embraced the role of the underdog of the restaurant business as he transgressed all its dominant ideals around 2010 by proposing alternative services, alternative wines, and an alternative (compared to New Nordic Cuisine) global food repertoire. Ironically, he did this so well that he even pleased the Michelin guide and received one star in 2012.

The Very Slow and Green Street

Jægersborggade has a distinct slowness to it. This is due in part to the near absence of cars and the calm passing of bikes on the bumpy pavement. The Christiania bike is very common. This is a cargo bike model with two front wheels and a large box between them used for children or groceries. Originally a hippie feature, it has now gone mainstream among middle-class families living in Copenhagen. Some shops even use these bikes for catering. For instance, the alternative ice cream shop Istid has a food-truck bike in front. The slow atmosphere is further enhanced by the many shop owners and visitors hanging out in front of their establishments. In front of a stationery shop, a child-size bench and table, equipped with free paper and drawing tools are on hand to entertain clients' offspring. The generous welfare policies of Copenhagen also allow Danish parents up to a year's parental leave. Also, an ordinary Danish working week is thirty-seven hours, so many parents pick up their children between 3:00 and 4:00 p.m. from school or day-care institutions. In Jægersborggade, parents on leave with their babies in strollers are visible during daytime hours and, from the early afternoon, children picked up from school in Christiania bikes are also in evidence.

In certain establishments, the slowness appears as an intentional strategy to distance themselves from a modern high-paced lifestyle. During one of my visits, I decided to have lunch at La Dispensa, an Italian (natural) wine and cheese shop with prepared food, such as sandwiches for lunch and cheese and charcuterie platters for dinner. They also sell biodynamic vegetables. The shop embraces the "slow food" spirit, an alternative culinary approach originating from Italy, and a far cry from mainstream narratives of greasy pizzas and heavy, sweet Amarone wines (Counihan 2019; Petrini 2007).

I ordered my food at the back. This room has an industrial look, like the former work space of a butcher's with white wall tiles and terrazzo flooring. In many of the shops and restaurants in the street, these back rooms, formerly backstage areas, are now included in the actual customer space. Allowing customers a peek behind the scenes reflects an ideal of openness and transparency. I chose a sandwich with goat cheese and bresaola, the only item on the menu at this hour. The person serving behind the counter pointed out that it would take fifteen minutes to prepare the sandwich, as there was already a customer ahead of me. I ordered a glass of wine as I waited. I was asked to go and get my glass from a counter behind me while my host lectured me on the biodynamic production of my orange wine. It had a sherry-like texture and a wild uneven herbal finish. A little too funky for my francophile palate. I sat down outside at an old wooden table with four colorful metal chairs in red and green. The paint was peeling off. After twenty minutes, I went back to check on my sandwich. It was still in progress. The only other guest was also waiting. I returned to my table. Ten minutes later, my host came out with the sandwich wrapped in brown paper. This was truly slow food. The service offered was expert knowledge and storytelling rather than simply catering to the hunger of customers. The taste of the bresaola and the goat cheese was very clean, but somewhat dominated by the bread, which was thick, flavorsome, and chewy. It obliged the eater to take the time to chew and enjoy it, contrary to the classic peanut butter and jelly sandwich.

La Dispensa also sells a small selection of biodynamic vegetables. Three dollars for a cucumber and palm kale equals ten dollars per kilo. It is manifest how many vendors of fruit and vegetables there are in Jægersborggade. Since my childhood in the 1980s, vegetable and fruit shops have vanished from the main streets in Denmark. They lost to supermarkets. Several new ethnic vegetable shops and bazaars often owned by people of Turkish or Middle Eastern origin have opened in different areas of Copenhagen. These shops are relatively cheap and they mostly source their products from outside Denmark. Organic produce is rarely a priority. In Copenhagen, it is a rarity to encounter a vegetable vendor with local and organic produce owned by an ethnic Dane. However, in Jægersborggade alone, we find three vegetable shops owned by ethnic Danes. The biggest is Løs Marked. Here, you find a great variety of vegetables, fruits, nuts, as well as oils, vinegars, and "chemical-free" soaps, etc. A central feature being that the shop does not provide bags or other packaging, so clients are obliged to bring their own. I was not aware of this and asked the owner for a plastic bag. The owner patiently explained the concept to me while a few other clients rolled their eyes. I clearly did not pass as a local.

Generally, vegetables are a big thing on Jægersborggade. Several establishments specialize in vegan versions of junk food. Havanas serves vegan ice cream. Plantepølsen is a vegan hotdog shop "with a kids-friendly menu and gourmet hotdogs for mom and dad." These examples highlight a broader negotiation in the street between sustainable eating and hedonism—and family friendliness.

The World's First Porridge Bar

While most types of shops in Jægersborggade might appear unique, similar kinds of shops and restaurants (natural wine bars, craft beer, vegan food) might be found in Global Brooklyn spots around the world. Also, they are all very active on Instagram and many are strongly connected internationally by this social medium. There are, however, examples of establishments you might not find elsewhere. The most striking is Grød, the world's first porridge bar. It opened in 2011 and was part of the early revitalization of the street as described by Puglisi. At that time, Lasse Skjønning Andersen, who conceived the idea, was just twenty-one years old. He dropped out of university to realize his dream of opening a porridge bar. Andersen describes his project as a way of going against the social expectations of finding stable career paths in the post-2008 financial crisis years. It was controversial, as many Danes are skeptical about eating porridge beyond breakfast. It is associated with a poor man's diet and, particularly among older generations, with the shortages resulting from the Second World War.

Andersen followed his passion for porridge and his dream like other millennials have done by upscaling old working-class jobs, food, and identities rather than pursuing traditional postindustrial middle-class careers (Ocejo 2017). So, he used his savings to engage in the project and as he went along he learned how to manage a culinary business. The restaurant on Jægersborggade is located in a basement with two small rooms for dining and a larger room at the back (again with a more industrial feeling), opening up to the kitchen with a counter where you order your porridge. The two spaces for eating at the front are neatly designed with new wooden tables accompanied by benches and chairs with colorful cushions. The establishment embraces a far more polished design than does La Dispensa or many of the other more retro *bricolage* style in the street's eateries.

The menu covers a range of porridges from traditional oat porridge to chia and risottos with peas and parmesan cheese. I chose the traditional oat porridge with crunchy almonds, caramel sauce, and small pieces of apple. The distinct texture of the topping and the balance of sweet and sour elevate the dish into a far more sophisticated experience than one might expect. It maintains, nonetheless, a distinct feeling of comfort food. No fermentation here: the challenge for most is certainly to embrace the concept of porridge as more than a quick breakfast or children's food. Also, this filling portion was priced at five dollars, which is very cheap at Jægersborggade where I paid four dollars for three onions at the vegetable stand just fifty meters down the street.

Grød also offers instant porridges, a cookbook, and different kinds of grains for home use. These are also sold in the upscale supermarket chain Irma, in the Danish 7-Eleven, and in 600 other shops in the Netherlands, according to Andersen. Also, Grød has opened restaurants in six other locations in Copenhagen and Aarhus, the second biggest city in Denmark. Thus, the question is whether this concept is becoming too big or too mainstream for Jægersborggade. Can Andersen with his porridge empire and Puglisi, who has also opened a series of restaurants across Nørrebro, still be considered underdogs or rebels? Or is the street entering a new phase as its culinary entrepreneurs become significant players in the food and restaurant business?

Some critics have noted that the street is not really as edgy as some like to think. "It is a very white street," remarks one of the inhabitants in a 2013 portrait in the male magazine *Euroman* (Nguyen 2013) This is still the case, and it is particularly striking as the street is situated in the middle of the very multicultural Nørrebro neighborhood where you are constantly surrounded by a multitude of languages, Middle Eastern bazaars, and people in non-Western dress. At Jægersborggade, you encounter quite a different expression of globalization, rooted in white cosmopolitanism and embodied in creative entrepreneurs and middle-class families on Christiania bikes.

Refshaleøen

The example of Jægersborggade illustrates how the city of Copenhagen has changed through systematic urban planning, particularly from the 1980s onward. The result is the massive gentrification of old working-class neighborhoods. The industrial areas have also moved away from the city center over the past decades. The meat-packing district is an obvious example. Here, the century-old buildings were preserved and used to create a new environment for creative businesses, galleries, restaurants, and nightclubs. Another example is the old Green Market in Valby. In this case, the industrial buildings from the mid-twentieth century were demolished and expensive apartments were built instead.

The area called Refshaleøen is situated at the Copenhagen waterfront somewhat isolated from the city center. This area housed the B&W shipyard between 1872 and 1996. It was an important industrial enterprise in Copenhagen with more than 8,000 employees at its high point. Upon closing, there was much debate about the future of this area comprising 500,000 square meters and including many old industrial buildings. Some argued that it could be the new Manhattan of Copenhagen with skyscrapers by the waterfront (Olesen 2002). Others maintained that it should become the new creative area of Copenhagen. Despite such debates, nothing much happened until 2010.

Amass

Around 2010, major events started to take place at Refshaleøen. In 2010, the heavy metal festival COPENHELL was held for the first time at Refshaleøen, and it has since returned annually. In 2011, the first version of the MAD (food) symposium organized by René Redzepi and his MAD Foundation was held in a gigantic tent on an abandoned field in the area. This was a major food event, which attracted a series of food celebrities such as David Chang, Massimo Bottura, Alex Atala, and numerous other members of the international food intelligentsia.

It was a few hundred meters from this venue that the chef Matt Orlando opened his restaurant Amass in 2013. Like Puglisi, Orlando had worked under Redzepi at NOMA and felt the time was ripe to open his own establishment. Amass is situated in an old warehouse formerly housing the tools used at the shipyard. The main room has a six-meter-high ceiling. The rough industrial feeling is maintained inside the dining room and rendered extra urban by graffiti.

Outside the restaurant, Amass has its own urban vegetable garden. In this way, the restaurant negotiates rough urban culture with a focus on New Nordic values like locality, freshness, and fondness for vegetables. However, like Puglisi, Orlando went to great lengths to distance himself from the dogmatism of the New Nordic movement with references to his American identity and the desire to serve great food in a more relaxed setting (Leer 2016: 8).

Amass is extremely concerned about sustainability, particularly evident in its dedication to reducing food waste and the restaurant's carbon footprint, "that is equal to half of what a normal restaurant operates at."[2] This is also evident in the menu where, for the past few years, the first dish has been "yesterday's bread," fried leftover bread from the previous day's serving. It is a particularly characteristic bread with fermented potatoes. The flavor is intense and the crust thick, a meal in itself with a variety of flavors and textures and quite the opposite to the more neutral French baguette.

Craft Beer

Compared to Jægersborggade, Refshaleøen is still a much wilder and much more unpolished postindustrial area. The old shipyard buildings are impressive, but their greatness is combined with a sense of chaos and decay. The old buildings are surrounded by open, unused areas and bumpy roads. The wind from the waterfront is often strong. This abandoned and untidy atmosphere also marks a distance from the orderliness, which dominates most of downtown Copenhagen only a few kilometers away.

After Amass opened, the interest in Refshaleøen rose. The number of creative spaces (art galleries and performance art spaces) grew and new food and drink venues followed. In 2017, the internationally acclaimed brewery Mikkeller opened another bar by the waterfront. Mikkeller is the most important Danish craft beer company. It was founded in 2006 by the high-school teacher Mikkel Borg Bjergsø and became an instant success in Denmark and, soon, also abroad. Mikkeller is particularly known for their experimental beers, using everything from Sichuan peppercorns to raspberries to push the limits of taste. Today it is a global brand sold in more than forty countries and it has become an iconic hipster beer. It is recognized by its ironic cartoon universe on the labels and its ambition to redefine beer.

At Refshaleøen, Mikkeller occupies an old factory by the waterfront called Baghaven, the back garden. Upon entering, you notice the big wooden beer barrels at the back. The beer menu is handwritten on a big blackboard behind the front and often changes in tandem with the small batches being used. The half-pint beers are served in small wine glasses and the pints in simpler "bodega glass" like the ones you would find in a British pub.[3] This also allows for two distinct styles of consumption: the more relaxed drinking-to-get-drunk approach or the more sophisticated I'm-a-beer-afficionado approach. I could observe both approaches as I visited. Some patrons were sniffing, gurgling the beer, and having very serious conversations about it, while others were just chatting and cracking jokes as you might see in any other bar. The majority of the hipster clientele did a little of both. First, paying attention to the taste, which was discussed, and then, after this initial and ceremonial tasting, they relaxed, talked, and drank freely. The clientele mimicked the brand's mix of fun and connoisseurship.

The uniqueness of Baghaven is underlined in a social media video with Ehren Schmidt, the chief brewer at Baghaven. Schmidt, a tall man with an imposing beard and a lumberjack shirt, explains how he has experimented with wild yeasts, which he found at Refshaleøen and the video shows him scraping a sample from a tree next to Baghaven.

Using beer brewing to relate to locality is also a key concept at the brew pub Broaden & Build opened in 2019 by Matt Orlando. The ambition is to infuse the local landscape into the beers and to work hard at discovering new, exciting beer and food combinations.[4] Like Amass, Broaden & Build occupies an old industrial space with an even higher ceiling of eight to ten meters, creating a warehouse atmosphere. This is probably as Global Brooklyn as you get in Copenhagen. The cement floor and bare walls intensify the postindustrial feeling. The brewery's big metal tanks are present throughout the pub. The tables are all communal with benches in a simple wooden design. There is graffiti on the walls in the style seen in Amass, similar to that of a cover of a fantasy novel. You order beer at a counter that is very similar to the one in Mikkeller and here, too, there are more than twenty different micro-brewed beers on tap in constant rotation. The staff is casual, but their beer expertise is eminent. Like at Mikkeller, the flavors are challenging and do not conform to the sensory norms of mass-produced beer. Intense bitterness, fruitiness, and sour undertones are very frequent. Unlike Mikkeller, Broaden & Build offers a variety of food in the form of tapas. These are mostly comfort foods with an edge, which surprise by their strong fermented flavors, as in the case of the fermented French fries.

Reffen

One of the most significant additions to the eating and drinking repertoire at Refshaleøen is the street food market Reffen. It opened in 2018 as a bigger version of the previous Copenhagen Street Food (2014–2017), closer to the center. The project was initiated by Jesper Julian Møller, who also owns an organic farm. The idea is to source most of the produce from his farm and other farms in the proximity of Copenhagen. The focus on sustainable practices is intense and Reffen obliged stallholders to follow the high ethical standards articulated in the mission statement for the street food market. The central dogma is "reduce and reuse."

The street food market is attractively connected to the postindustrial architecture at Refshaleøen by the use of cargo containers as "food trucks." There are also indoor communal spaces with a *bricolage* of old furniture, but in the summertime, the most obvious choice is to sit by the waterfront or at the faux beach in the center of the market with beach chairs. The small skater park between the "beach" and the waterfront can be freely used by all—clients or otherwise.

The offering covers a wide range of street food from all over the world, including elements new to the Danish market like Nepalese and Peruvian cuisines. However, the assortment of foods and flavors appear more comforting and less innovative and challenging than at Broaden & Build or at the porridge bar in Jægersborggade. As the review in the Copenhagen-based newspaper *Politiken* noted, the barbecue smoke from the many burger and kebab stalls dominates the sensorium. Barbecue smoke might not be the best expression of the ethical profile outlined in the mission. The review questions whether the Copenhagen foodies who are used to more refined, lighter, and more vegetable-based foods will embrace Reffen and its global junk/street food.

This remark underlines a potential dilemma. Many of the vendors might go for the "safe" choices like burgers, pasta, and kebab to please a larger clientele and certainly also most of the tourists (I notice that the tourists outnumber the locals, particularly in the summer months). This might render the establishment less attractive to the critical mass of the Copenhagen foodies who have been spoiled with culinary innovation for more than a decade.

Chic Brooklyn and Rough Brooklyn

The Global Brooklyn spaces in Copenhagen bear many resemblances to the sensibility outlined in the introduction to this book, with a nostalgia for postindustrial and "authentic" working-class spaces and the intellectualized search for edgy flavors and ethical consumption, from the vegetable focus on Jægersborggade to the "reduce and reuse" ethos of Reffen.

Nonetheless, the case of Copenhagen could nuance some of the aspects of this mythology. Firstly, Global Brooklyn seems to be antithetical to fine dining. In the case of Copenhagen, there is a complicated relationship between, on the one side, NOMA and New Nordic and, on the other, the Global Brooklyn momentum in Copenhagen. It is noticeable that the two "defecting" NOMA chefs have played a central role in bringing legitimacy to the new gentrified areas. Both Amass and Relæ are negotiating a kind of relaxed, but high-quality dining in a Global Brooklyn-style setting. Puglisi's and Orlando's entrepreneurial dreams were to do something distinctly different from NOMA and these ambitions have facilitated the development of a series of culinary Global Brooklyn-style entrepreneurs in Jægersborggade and at Refshaleøen. One might say that the backlash against New Nordic has assisted the development of Global Brooklyn in Copenhagen. Furthermore, the international hype of the New Nordic has attracted more foodie tourists to the city. The tourists obviously do not eat all their meals at NOMA and are thus a potential major source of income for the many Global Brooklyn-style "indie places." Also, many of the people running these establishments are not Danish, but came to Copenhagen because of the culinary innovation of NOMA, Orlando being a case in point.

Secondly, the contrast between Jægersborggade and Reffen also forces us to nuance Global Brooklyn and talk about different versions of this imagery. Both areas present the ethical mindset, entrepreneurial spirit, and the challenging food concepts in a casual, retro, postindustrial decor. Jægersborggade, however, appears much more polished, ethnically homogenous, and friendly to the urban, Danish middle-class population and their children than Refshaleøen. The latter is more chaotic, cosmopolitan, and essentially postindustrial. We could distinguish between a chic version of Global Brooklyn (Jægersborggade) and a rougher one (Refshaleøen). However, this distinction is not strictly bound to locations. For instance, at Jægersborggade, there are both rougher and more chic establishments. La Dispensa is very casual and unpolished in terms of interior, service, and concept. In contrast, Grød is much sharper and more streamlined with a clear concept (growing into a franchise), a stylish interior, and a much more professional service.

A theoretical contribution of this chapter might then be a call to distinguish between different versions of the Global Brooklyn imagery and to think about how these coexist between alternative and mainstream food cultures. When does the alternative Global Brooklyn stop and when does mainstream culture take over? Businesses turning into franchises (like Mikkeller and Grød) started out as rebellious entrepreneurs gambling a safe career on the food craft they love. As their businesses expand and turn into global brands and chains, how can they claim to be authentic or to represent true alternatives to mainstream capitalist culture? Are they dropping out of Global Brooklyn? Or rather, are they expanding its concept, opening it to mainstream culture? The inclusion, in a Danish context, of families in Global Brooklyn could be seen as a first step toward a mainstreaming or IKEA-fication.

Notes

1   Due to space limitations, I leave out Kødbyen (the former meat-packing district) and the nearby Vesterbro, which are also highly important parts of the Global Brooklyn and hipster food scene of Copenhagen, see also Lapina and Leer 2016.

2 https://www.forbes.com/sites/christinaliao/2019/01/23/matt-orlando-noma-for
   mer-head-chef-opening-broaden-and-build-brewery/#7722efde4f98
3 See the differences between the bodega glass and the Henry and Sally Glass on this
   link: https://shop.mikkeller.dk/collections/glassware
4 The Danish newspaper *Politiken*, January 11, 2019.

References

Andreassen, R. (2014), "The search for the white Nordic: Analysis of the contemporary New Nordic Kitchen and former race science," *Social Identities*, 20 (6): 438–51.

Counihan, C. (2004), *Around the Tuscan Table: Food, Family, and Gender in Twentieth-Century Florence*. New York: Routledge.

Counihan, C. (2019), *Italian Food Activism in Urban Sardinia: Place, Taste, and Community*. London: Bloomsbury.

Haraldsdóttir, L., and Gunnarsdóttir, G. (2014), "Pure, fresh and simple: 'Spicing up' the New Nordic Cuisine." In L. Jolliffe (ed.), *Spices and Tourism: Destinations, Attractions and Cuisines*, 169–81. Buffalo: Channel View Publication.

Jensen, T. (2014), "Pork, beer, and margarine. Danish food consumption 1900–2000: National characteristics and common nordic traits," *Food and History*, 12 (2): 3–37.

Jönsson, H. (2013), "The road to the new nordiccuisine." In P.Lysaght (ed.), *The Return of Traditional Food*, 53–67. Lund: Lund University Studies.

Krogsgaard, M., and Madsen, R. (2010), "JægersborggadeogKronborggade: 2 gaderpåNørrebro under forandring," *Geografiskorientering: tidsskrift for Geografforbundet*, 40 (6): 6–14.

Lapiņa, L., and Leer, J. (2016), "Carnivorous heterotopias: Gender, nostalgia and hipsterness in the Copenhagen meat scene," *Norma*, 11 (2): 89–109.

Leer, J. (2016), "The rise and fall of the New Nordic Cuisine," *Journal of Aesthetics & Culture*, 8 (1): 1–17.

Lonely Planet. (2019), "Copenhagen." Available online: https://www.lonelyplanet.com/d enmark/copenhagen (accessed September 19, 2019).

Lisberg, H., and Thorsen, N. (1993), "Gaden, der erelsketoghadet," *Politiken*, February 28.

Neuman, N., and Leer, J. (2018), "Nordic cuisine but national identities: Nordic cuisines and the gastronationalist projects of Denmark and Sweden," *Anthropology of food*, 13. https://journals.openedition.org/aof/8723

Nguyen, K. (2013), "Jægersborggade Blues," *Euroman*, 233:76–9.

Ocejo, R. E. (2017), *Masters of Craft: Old Jobs in the New Urban Economy*. Princeton: Princeton University Press.

Olesen, P. (2002). "Københavns Venedigeller Manhatten," *Berlingske Tidende*, January 23.

Petrini, C. (2007), *Slow Food Nation: Why Our Food Should Be Good, Clean, and Fair*. New York: Rizzoli Ex Libris.

Puglisi, C. (2014), *Relæ: A Book of Ideas*. London: Ten Speed Press.

Puglisi, C. (2013). "'Move Your Gut or be Gutted' at the MAD symposium 2013." Available online: https://www.youtube.com/watch?v=28-eHej-_V0

Skårup, B. (2013), "The new Nordic diet and Danish food culture." In P. Lysaght (ed.), *The Return of Traditional Food*, 33–42. Lund: Lund University Studies.

4

Global Paris

Between Terroir and *Hamburgés*

Susan Taylor-Leduc

Mahaut Landaz, a journalist at the weekly *Nouvel Observateur* magazine, recently declared that the "Brooklynization" of Paris was no longer à la mode. The red brick walls and industrial lighting that had invaded French restaurants and co-working spaces were now considered *ringard*, a negative slang word connoting kitsch, a condemnation that implied the imminent demise of a global phenomenon (Landaz 2019). Landaz argued that the ubiquitous exposed brick no longer signaled gentrification nor hipster culture, but a gesture toward upscale marketing. Rethinking Landaz's prediction in the culinary arena suggests another interpretation: chefs, restaurateurs, and tastemakers have assimilated Global Brooklyn and made it French.

French gastro-nationalism has been a cultural identifier for centuries, one that not only defined Frenchness but has inspired millions of tourists to travel to France to seek food experiences (DeSoucey 2010: 432–55; Ferguson 2004: 165–74). This form of gastro-nationalism is most closely associated with chef Auguste Escoffier (1846–935), who organized his kitchen into a brigade system so that he could design his dishes to be included in multicourse meals (Trubek 2000: 10–26). Escoffier's meals and methods were conceived for fine dining experiences where consumers were expected to understand his copious menus, order wines paired with food, and be able to use tableware and glasses properly (Escoffier 1903). This etiquette became so identified with French cuisine that it promulgated a form of internationally recognized culinary cultural capital.

French haute cuisine is still alive and well in Paris at palace hotels and Michelin-starred restaurants. Nonetheless, today's tastemakers have contributed to a reassessment of the viability of haute cuisine as the primary signifier for

French food. Franco-American journalist Lindsey Tramuta, in her book *The New Paris*, has documented how the development of bistronomy in the 1990s opened a breach in the haute-cuisine hegemony over the Parisian food scene. Inspired by chef Yves Camdeborde, professional chefs left the high-end world of gastronomic dining and revisited traditional bistro dishes (hearty one-course meals such as pot-au-feu and regional stews), adding new flavors and ingredients driven by market availability. These passionate chefs attracted savvy consumers interested in experimenting with new food combinations at more modest price points. Tramuta posits that following the financial crisis of 2008, a new generation of DIY chefs spearheaded a fresh incarnation of bistronomy, explicitly joined to worldwide concerns about sustainability (Tramuta 2017: 33–41; Parasecoli 2019).

This chapter proposes that some of the features of Global Brooklyn converge in what Tramuta has defined as the New Paris food culture. Both movements share concerns about sustainability, food quality, and a shared culinary community. However, while Global Brooklyn elsewhere highlights restaurant design and active social media engagements to promote new venues as connected communities, in France the appeal to an international network is less important than the respect for artisanal practices and local products from the French terroir. Chefs may bring international culinary tastes and experiences to France, but ultimately it is their ability to fuse their skills to French dishes and products that attract foodies, millennials, and tourists avid for new tastes and sensory experiences, but still wanting to identify with French cooking and foodstuffs uniquely available in France.

In the early twentieth century, the idea of terroir was supported by the development of legislation known as the *appellations d'origine contrôlées*, a system of intellectual property on distinctive wine and food products put in place in the 1920s and institutionalized in 1935 (Trubek 2008: 21–31). These laws first appeared following the devastation of the French food supply system in the First World War, but have morphed into a means to glorify rural France and the richness of French soil as source for exceptional foods (Trubek 2008: 21). In the 1930s, and again after the Second World War, food writers from both the left and the right of the political spectrum promoted culinary chauvinism, an affirmation of national identity that exists outside of history as rooted in gastronomy, its wines, and terroir (Parkhurst Ferguson 2010: 105). As Amy Trubek has argued, terroir became a means to symbolize how "ingredients and dishes came to represent their regions, ultimately guaranteeing their permanence, for they came to signify more than a dish using locally available ingredients (bouillabaisse in and around Marseille, cassoulet in and around Carcassonne), but also to represent the taste of that place, wherever the dish may be consumed" (Trubek 2008: 38).

Trubek elucidates how terroir became politicized:

> These men and women (aka tastemakers) observed their world and decided to champion certain practices (small farms, regional dishes) and values (tradition, local taste) in order to make sure that they did not disappear. Their cultural and economic investments made the French word for soil signify so much: a sensibility, a mode of discernment, a philosophy of practice, and an analytic category. What they said may have embraced the timeless and essential notion of mother Earth, but what they did was to create a vision of agrarian rural France and convincingly put it in people's mouths. (Trubek 2008: 21)

This vision of an eternal agrarian France that will nourish the nation has long-standing appeal on both ends of the political spectrum (Davis and McBride 2008; Parkhurst Ferguson 2010: 104–6).

After the Second World War, the French governments continued to adopt programs to protect terroir, considering that foods and artisanal practices constituted a national patrimony. For example, under the socialist government of François Mitterrand, Jack Lang, the then minister of culture, established a national commission in 1990, the Conseil National des Arts Culinaires, which turned to terroir as a means to promote French culinary culture and transmit it from one generation to the next (Tomasik 2007: 238–42). Although disbanded ten years later, as a government subsidized think-tank, the CNAC contributed to the development of a program for primary schoolchildren, the *semaine du goût*, to help develop "French" palates and a taste for a wide range of foods from different regions (terroirs). The program is so successful that it continues thirty years later.[1] Today's incarnation of the *semaine du goût* has been re-oriented to ensure that all children have access to quality food experiences, now expanded from primary schools to university campuses. Another example of government support for terroir is the development of tourism to encourage visits to sites where specific foods are produced. A label was created for travel destinations: "sites remarquables du goût," or "remarkable sites of taste."[2] By far the most important effort to secure French terroir was the inclusion of local products in the "French gastronomic meal" included in the UNESCO intangible immaterial culture list in 2010. One of the criteria from the description of the French meal for UNESCO hinges upon a recognition that home chefs would "select preferably local products (and wines) whose flavors go well together."[3]

Generations of French diners, from all social classes, are expected to appreciate the products from the terroir. It is precisely the intersection of terroir with Global Brooklyn that poses interesting questions: how does cultivation of local products contribute (or not) to authenticity? Can the traditions of terroir and the fact that these products are equally integrated in haute cuisine as signs of luxury accommodate the values and practices of Global Brooklyn? How does terroir influence regional cuisines that have contributed to both local and national culinary identity, when faced with the international influence of Global Brooklyn? (Csergo 2016: 500–15; Trubek 2008: 33–6). This chapter offers three case studies to question how Global Brooklyn has been assimilated to Paris, recognizing that these influences may have different outcomes in Lille, Marseille, or Lyon, where regional practices continue to influence local pride of place.

The first case study focuses on hamburgers, examining how the most iconic American dish set the stage for appreciation of Global Brooklyn in Paris. The second study looks at restaurant design in the Batignolles district, a section of the seventeenth *arrondissement* that actually has no tourist attractions or historic monuments, but is now home to a Texan barbecue restaurant, a specialty coffee shop, and a workshop-brewery. For the merchants and restaurateurs in the Batignolles, an area comparable to the Park Slope section of Brooklyn, appealing to Global Brooklyn has been tantamount to the gentrification of the neighborhood. The third case study examines how the Global Brooklyn aesthetic has appealed to one of France's most successful celebrity chefs, Alain Ducasse. While the Ducasse empire includes Michelin-starred restaurants, palace hotels, and cooking schools that exemplify French luxury branding, Ducasse's latest venture, his Manufactures (literally "factories") for coffee and chocolate, reveals why Global Brooklyn has yet to become *ringard*.

Hamburgers/*Hamburgés*

While Global Brooklyn is considered an international phenomenon, the reference to an American place writ large speaks to the history of Franco-American culinary exchanges. Since the end of the Second World War, the simultaneous attraction and disdain for American food in France has been fraught with political overtones and economic anxieties, coalescing around the aggressive marketing of transnational corporations such as Starbucks and McDonald's, whose commercial ventures have invested in the cityscapes and suburbs of France (Fischler 2016: 530–47; Henault and Mitchell 2018:

274–82). This history does not need to be rehearsed here, but it is important to acknowledge how the assimilation of American foods, most conspicuously the hamburger, predates Global Brooklyn but nonetheless sets precedents that both facilitated and tainted the reception of American foodways. Most notably, the battle against the omnipresence of McDonald's in France has been framed as a David versus Goliath or Asterix versus the Romans story recast for food culture. In this food fight José Bové, a charismatic farmer from the Ariège region, spearheaded an attempt to prevent the installation of a McDonald's to protest an American tax on Roquefort cheese in 1999. Twenty years later, Bové's protest is considered a factor in urging McDonald's to outsource to local producers, so that the firm is now highlighting its contribution to the French economy and job-training programs rather than its significance as an American behemoth (Midi Libre 2019).

No longer a sign of "McDonaldization," one hamburger restaurant franchise invites its clients to start a culinary romance with the newly renamed *hamburgés*.[4] Another franchise proposes that creating the hamburger is an artisanal craft, with chefs "curating collections" of hamburger dishes that change with every season.[5] For hamburger franchises, Frenchifying the hamburger is a savvy advertising tool that both supports and subverts stereotypes of American cuisine. These so-called French "improvements" upon the industrialized fast food, by adding recognizably French ingredients to the sandwich, have not substantially changed the signature dish but did increase prices and profits.

By contrast, Kirsten Frederick, an American chef in Paris trained at the prestigious Ferrandi cooking school, single-handedly re-introduced the specifically American hamburger to Paris when in 2011 she opened the food truck Camion qui Fume (Tramuta 2017: 50–9).[6] Frederick's stroke of genius was to offer quality food at a reasonable price to an increasingly nomadic crowd. Prior to her food truck, street food in France was considered an affront to both the high-end and the low-brow French luncheon culture, be it the three-hour gastronomic business lunch or the pop-up street vendor grilling merguez sausages of questionable quality. Eating out-of-doors, even standing up at a food truck, soon replaced the stodgy *jambon-beurre* sandwich devoured in front of a computer. Focusing on quality products and deploying American recipes designed to evoke the flavors of an authentic American food experience, Frederick's hamburgers offered a real alternative to the Frenchified *hamburgés* franchises.

Frederick helped pioneer a new concept of consumption (*bien/pas cher/ bon*, that roughly translates into "good ingredients/not expensive/delicious") that benefited from a long-standing appreciation of American entrepreneurial success stories in France.[7] For foodie DIY entrepreneurs, food trucks are relatively cheaper to open and operate than restaurants, reflecting the flexibility and practicality long associated with American business and now essential to the post-2008 gig economy.[8] Eleven food trucks opened in Paris in 2013. Today the city sponsors over a hundred food trucks at several locations several times a week.[9] At first reluctant, the city of Paris now endorses food trucks to promote a dynamic Parisian food scene.[10] This endorsement was not lost on the restaurant community. Celebrity chef Thierry Marx, who founded an association, *Street food en movement*, jumped in to fill the gap creating links between the French food industries, restaurants, and public authorities. The association belies some of the anxieties provoked by the success of street food to owners of cafés, brasseries, bistros, and restaurants that often have terrace seating open to the street. Most importantly, the charter of the association suggests that the movement was not inspired by American innovation, foods, and foodways, but rather promotes "a valorization of terroir products and of culinary heritage"—in other words, making street food French.[11]

From the perspective of Global Brooklyn, the development of a duck (*canard*) burger at the restaurant Canard Street disrupts the imitation of an American dish with the creation of a new French alternative.[12] For Francophiles interested in French regional specialties, duck is particularly associated with the terroir of the Southwest of France. Raising ducks is linked to ancestral farming practices, regional recipes such as *confit de canard* and the luxury dish *foie gras*. The launch of a "*canard*" restaurant, deliberately open to the street and with a visible working kitchen space, does not necessarily connote the Southwest or fine dining. Instead the interior—parquet floors, red and black painted walls, some potted plants, bar stools, and industrial furniture—suggests Global Brooklyn, a casual space for consuming an exceptional product.

Founder Gregoire De Scorbiac is particularly interested in street food. On a trip to Hong Kong, impressed with the presentation of duck as an accessible commodity, he mused about the special attitude of the French toward duck. Upon his return to Lille, he teamed with Nicolas Droulat, his fellow student at their business school, to transform the traditional French ways of eating duck connected to terroir, so that it could be consumed as street food. With the support of the mayor of Lille, the student-entrepreneurs opened a "corner stand" in a municipally supported farmers' market, and developed duck dishes: *canard* hamburgers, salads, sandwiches, and raw duck tartare. In addition, the mayor required that they sell products from their stand, which extended the range of duck dishes to include packaged *foie gras*, duck sausages, and canned *confit de canard*. They have continued to promote the storefront (*épicerie*) as part of the Canard Street restaurant experience although this was not their original intention. The market corner served a local craft beer from Lille—from the Brasserie Celestin, which has existed since 1747, signaling their attachment to the northern region.

One year later, the team expanded from Lille to Paris and opened two restaurants geared primarily toward a lunch-time business crowd who would be attracted to duck, but appreciate the affordable prices. They proudly broadcast their respect for the French duck: there are ducks on their logo and graphic design, and duck decoys are artfully arranged as sculptures on the wall. On their website and menus, they highlight the quality and traceability of all their products. While animal rights activists may be repelled by duck consumption, the owners work with local producers to source their restaurants and oversee the transformation of duck into hamburgers, sausages, and confit, emphasizing the versatility of the product. Moreover, the potatoes, again locally sourced from northern France, are fried in duck fat, enhancing the taste and demonstrating the desire to profit from every aspect of the duck and minimize waste. They continue to promote Celestin beer in bottles and draft in Paris, linking both the northern and the southwestern terroirs.

From Hong Kong to Lille to Paris, these students turned entrepreneurs have transformed terroir into Global Brooklyn experiences. Customers are invited to admire the product, watch an open kitchen, and sit at shared counter top tables surrounded by the brick-exposed walls that recall New York. Most importantly, the customers are enjoying a recognizably French dish. The hamburger is not an imitation of an American model; it has become French. Significantly, the restaurant does not promote an online community, while posting on Instagram to highlight their product: it is the product that attracts and unifies the clients. Moreover, Canard Street shows how innovation in one region—Lille—joined with another region—the Southwest—has migrated to Paris, the international center where both these regions can be linked to Global Brooklyn.

Global Brooklyn in the Batignolles

Founded in 1830 under King Charles X, the commune of Batignolles-Monceau, located to the northwest of the city center, was a small village (Bellhoste, Lohr, and Smith 2001: 338–52; Rouleau 1985: 303–34). At the moment of its annexation to the city in 1860, the Prefect of the Seine, Baron Haussmann, split the village effectively in two—the Plaine Monceau, near the former eighteenth-century garden, now the Parc Monceau—and the more industrial, working-class Batignolles located near the railroads. Today, the industrial history of the Batignolles, where workers constructed railroad cars, has effectively disappeared with the creation of the park Martin Luther King, which was built over the disaffected freight zone.

The gentrification in the Batignolles is considered anything but *ringarde:* the former working-class village, where the architecture is scaled to modest apartments and family businesses, is now recognizably hip or *branché,* where upscale franchises but not yet couture boutiques appeal to fashion-conscious millennials. Home to a city market, one that was originally designed by Victor Baltard, modeled after the iron and glass markets at les Halles and likewise torn down in the 1970s, the neighborhood still has a market at its culinary center. In addition, at the northern limit of the *arrondissement,* the Boulevard de Batignolles is home to a bi-weekly organic food market. While the city market remains affordable and attracts local clientele, the organic market draws shoppers from around the city who can afford its high quality and prices. Boutique supermarkets of sustainable products have established themselves on the boulevard.

For aspiring restaurateurs, until recently the Batignolles offered affordable rents within city limits. Two recognizably Global Brooklyn venues—Dose, a specialty coffee shop, and Melt, an American barbecue—have asserted their place in the neighborhood. The owners are determined to establish neighborhood identities and their success has attracted a growing number of neo-bistros appealing to a clientele that eschews fine dining in favor of more informal gatherings.

Dose

In November of 2013 Gregoire Reverse, during an internship at the Terra Nera Coffee shop in Camden Town, London, became enamored with specialty coffee beans and the slow coffee movement. Upon his return to Paris, he was determined to open a quality coffee shop and launched an espresso bar, geared to students and tourists, in the trendy fifth *arrondissement.*[13] Reverse was not alone, but part of a wider trend underway since 2005 when Gloria Montenegro, a former ambassador from Guatemala, opened the Caféothèque, a coffee shop, roastery, and coffee school, and introduced "cupping" to the center of café society (Tramuta 2017: 82–115). While meeting for coffee in a café is a French national pastime, Montenegro helped introduce entrepreneurs to roasting to better quality beans (Tramuta 2017: 86–8).[14] Reverse was part of this growing trend, and working with his associates, Jean-Baptiste and Mathilde Perez, decided to open Dose in the Batignolles.

The Batignolles incarnation of Dose opened in 2016 and was built upon the experience learned from the espresso bar, which was originally billed as having a café as a "daily dose," to offer more seating and a variety of roasted blends. Juliette Champenoise oversees the selection and roasting of beans at the collaborative *torréfacteur* shop called Beans on Fire, keeping the focus on quality specialty coffee. While seeking to offer a quality cup, the shared tables promote hospitality and low-key service.[15] Designed by the Dose team, the interior is a Scandinavian riff on Global Brooklyn, with wooden tables and floors, shared seating, and a small open kitchen. The café is open from 8:00 a.m. to 6:00 p.m., Wi-Fi is only offered on weekdays, and although some consider it a co-working space, the team ensures that the café is open to an intergenerational crowd. The menu is limited but includes some Anglo-Saxon references: scones, cookies, carrot cake, and avocado toasts are decidedly not French; but the food is locally sourced and prepared in the neighborhood. None of the associates were trained as chefs.

Dose offers a recognizable Global Brooklyn interior design, but the owners continue to focus on the café as a place for drinking quality coffee. Dose promotes Instagram but relies on its friendly service to welcome a local crowd. They have expanded their online presence to sell their blended coffee to a wider audience. They have designed a sleek packaging and logo, created by the graphic and tattoo artist Jean André, for their roasted blends that they hope clients can pick up on the premises, keeping their connection to the café experience. Dose certainly followed the wave of the slow café movement, but it has also embraced Global Brooklyn to rethink the making of place, promoting hospitality, and informal service as a neighborhood establishment for an intergenerational crowd.

Melt

While Dose hovers between café and co-working space, the owners of Melt were emphatically dedicated to the restaurant business.[16] As aspiring entrepreneurs, looking for a niche market, they discovered barbecue in New York while on student exchange programs. Jean Ganizate and Antoine Martinez both graduated from the Swiss Glion program in hospitality and management in June of 2013. They left for New York to begin their careers in restaurant management.

Living in the East Village, they discovered a Texas barbecue restaurant, Mighty Quinns, a franchise where they were exposed to another American tradition: counter service. Enthralled with barbecue and convinced the concept did not exist in France, they embarked on a road trip to the barbecue belt—Houston, Dallas, Austin, and Galveston—to learn about that tradition. They invited Jeffrey Howard, a pit master from one of the best barbecue restaurants in Dallas, to France to collaborate on their project for the next two years. Although not the first barbecue restaurant in Paris, in 2016 they opened Melt in Oberkampf in the eleventh *arrondissement*. This restaurant in many ways has been a training ground for the Batignolles venue, opened in December 2017 (Tramuta 2017: 61–2).

When they moved to the Batignolles in 2017, Jean Ganizate and a third partner, Paul Loiseleur, expanded Texas barbecue to include sauces from Morocco, Thailand, and Mexico. After three years elaborating the menu they proudly declared their mission: barbeque is respect for cuts of meat, sauces, and a long wood-smoking process. This appreciation for meat is keyed to a long-standing French recognition of quality beef and butchers' ability to provide special cuts that are part of artisanal knowledge and practice. Clients can see the kitchen behind the counter, watch the assemblage of dishes, and most importantly smell the aromas of barbecue. The emphasis on smoking meat as an art form was clearly promoted: the waiters who facilitate the counter service wear T-shirts that declare the pleasure of eating smoked meat every day. At the Batignolles location, the aroma from the kitchen and the wood for the stoves are key to its success: it is a sensual indulgence and a novel taste sensation. Critical to the success of both restaurants is the counter service: it reduced costs but also suggested a uniquely American experience made to order. At both restaurants, one orders at the counter and takes a tray to the table. The prices are moderate, easily visible at the counter, and their block lettering recalls American diners.

Dose and Melt are not the only recognizable Global Brooklyn venues in the Batignolles. The Houblonneurs, which resembles a pop-up store, teaches groups how to make artisanal beer.[17] This is not a place to buy craft beer, but rather an atelier experience where you take home the beers you have made (20 liters) after a shared workshop that combines tasting and the enjoyment of a DIY experience. These venues in the Batignolles reflect the long-standing relevance of the working-class imaginary in the neighborhood where manual labor has retained positive associations and the local population remains predisposed to appreciate Global Brooklyn work ethics.

Prior to the arrival of either Dose or Melt, the Belgian chef, Wim van Gorp, who was trained under Alain Ducasse and Jean George Vongerichten, opened a neo-bistro called Comme chez Maman in the Batignolles in 2011.[18] Following paths forged by the first generation of New Paris chefs, van Gorp opted out of the gastronomic experience for more home style cooking. He revisited traditional bistro dishes and went so far as to bring cast iron pots with steaming stews directly to the table. The slate menus, informal service, and exposed brick resonated with the Global Brooklyn aesthetic. Van Gorp expanded his business on the same street with Wim à Table in 2018, a wine bar where his inventive cuisine is served in the form of a tasting menu in small shared plates.[19] At Wim à Table clients sit on stools at shared wooden tables, surrounded by brick walls. The menus written on slate reflect daily changing ingredients and an extensive wine list.

The attraction of the Batignolles as a place for Global Brooklyn and New Paris establishments was confirmed by the opening of the Le Petit Boutary, in 2019, an outpost of the restaurant and caviar purveyor at the Maison Boutary on the Left Bank.[20] While the Left Bank establishment is modern and gastronomic, in the Batignolles the design of the Petit Boutary is resolutely Global Brooklyn: exposed brick walls, wooden tables and chairs, tile flooring, and unadorned lighting, all signaling its informal atmosphere. The fusion chef, Jay Wook Hur, offers a surprise menu: clients choose the number of dishes without knowing the ingredients, expecting that the chef will create dishes inspired by market availability. Chef Wook Hur may or may not include the signature Boutary dish, caviar, which offers a twist on the connection to terroir. Caviar is not a regional dish, but it has been made French thanks to the Boutary family who has been importing and marketing it since 1888. At the Petit Boutary, an international chef brings his Asian expertise and surprises his clients. But savvy consumers are ready to pay for their savoir faire and creativity in a carefully designed space reflecting global aesthetics, mitigated by the Boutary ties to terroir, which anchor the experience in French culinary tradition.

Is the Batignolles the new Brooklyn of Paris? Yes and no. Certainly the constellation of restaurants in the Batignolles demonstrates how the New Paris/ Global Brooklyn venues are mutually influencing places, appealing to an urbane clientele who expect sustainable practices, choice ingredients, and authentic culinary experiences at relatively modest prices. Most importantly, the exposed brick design and informal service connect these restaurants to one another and to Global Brooklyn, affording their hip but not necessarily hipster diners the ability to belong to an international network while ultimately maintaining French gastro-nationalism.

Les Manufactures Alain Ducasse

The third case study focuses less on how Global Brooklyn has been integrated into Paris, but examines how a French luxury brand recycled some of the Global Brooklyn design and ethics in order to promote French savoir faire and cultural identity. Alain Ducasse's food empire is a testimony to his extraordinary talents as a chef and entrepreneur, a modern Escoffier, attuned to an international audience searching to claim cultural authority by participating in his cooking and restaurant experiences. Ducasse turned to two products—chocolate and coffee, both decidedly not grown on the French terroir—and reworked them to incarnate French savoir faire. Once branded and repackaged as Ducasse products, they enter the international market surfing on the Global Brooklyn trends through stores called Manufactures, literally "factories" in French.

Situated in the eleventh *arrondissement*, both stores are meant to evoke the working-class neighborhood. Located behind stone facades, the refurbished iron and glass ateliers remind us of the industrialization of nineteenth-century Paris. The retail spaces are designed to showcase labor: at the coffee manufacture, vintage machines are displayed in the shops, juxtaposed with state-of-the-art machinery necessary to the selection and refinement of coffee beans. Beyond glass partitions, client-consumers watch apprentice workers verify the quality of the coffee beans or the transformation of chocolate into bars (*tablettes*) or candies (*bonbons*). Here the fabrication of chocolate and coffee is given pride of place with an emphasis on artisanal practices, gestures, and time, respecting schedules for roasting coffee beans melting and fashioning chocolate. At the same time, these glass windows partitions separate the world of luxury consumption from that of production.

For both Manufactures, Ducasse asserts his culinary talents to justify his expansion into marketing. According to the company's press releases, Ducasse's interest for coffee sprung from his preoccupations as a chef: his curiosity to discover and promote coffee as a gustatory experience that concludes the haute-cuisine meal at Ducasse restaurants. Ducasse equates roasting in the kitchen to roasting coffee beans, again asserting his mastery of culinary techniques. "Just like in the kitchen, bean roasting is a cooking process: it is a highly technical art for which a lot of sensibility and a little bit of magic are necessary."[21] Similarly, chocolate is considered a noble product that requires the skills of a pastry chef to exercise their savoir faire transforming a rare crop into a gastronomic delight.

At the Manufacture café, one is invited to watch the selection of the beans and to admire the baristas, their gestures and service, but not to share the experience or to linger in the café, while savoring the finely selected blends of Ducasse coffee.

For Ducasse, the café becomes a place linked to fine dining, with the "degustation" of coffee as a ceremonial tasting, similar to wine in terminology and praxis. After the "tasting," the impeccable high-design packaging encourages the purchase of coffee, either in grains or ground. Finally, Ducasse has commissioned a series of products for making and serving coffee, assuring that the *haute-cuisine* experience can be extended to the home.

The appropriation of Global Brooklyn has now gone full circle, coffee and chocolate are culinary art forms, the packaged products marketed internationally as signs of French savoir faire and culinary nationalism—to be consumed by a Global-Brooklyn-ready crowd. The crossover from DIY to luxury branding needs to be considered not only as a contemporary marketing tool, but it also suggests how Global Brooklyn can be repackaged as French, reentering the market as signifier of French savoir faire, thus revealing how Frenchness continues to serve as a sign of cultural identity even when the product is no longer grown on the French terroir.

Conclusions

This review of Global Brooklyn in Paris has attempted to analyze a decidedly fluid international phenomenon in the French context. The discussion of the hamburger examined how Global Brooklyn entered a culinary landscape already steeped in Franco-American culinary exchanges. The review of French gastro-nationalism, in particular the celebration of terroir, has demonstrated that since the 1920s, French consumers of all classes have been educated to appreciate specific French foods as signs of authenticity and taste. The desire to preserve terroir has segued into twenty-first-century concerns with sustainability that join the essence of French cuisine to Global Brooklyn. At the same time, both established and DIY French chefs have created a new food scene in Paris, and in so doing, they assimilated the postindustrial design, informal service, and sustainable ethics, while asserting their Frenchness onto the Global Brooklyn networks.

Notes

1   https://www.legout.com/la-semaine-du-gout
2   https://sitesremarquablesdugout.com

3 https://ich.unesco.org/en/RL/gastronomic-meal-of-the-french-00437#identification

4 https://bigfernand.com/

5 http://lartisanduburger.com/gb/chefs-artisans/

6 https://lecamionquifume.com/histoire/

7 TV France 3 October 14, 2013, https://youtu.be/ia8twc2WcUU

8 TV 5 Comment Ouvrirun food truck? February 18, 2015. https://youtu.be/cD018UVj-XI

9 https://tttruck.com/

10 https://en.parisinfo.com/where-to-eat-in-paris/info/guides/where-to-eat-what-you-feel-like-eating-in-paris/street-food-in-paris/street-food-in-paris

11 https://www.streetfoodenmouvement.fr/charte-qualite/

12 https://www.canard-street.fr/

13 https://www.dose.paris/content/8-nos-cafes-de-quartier

14 https://parisbymouth.com/percolating-in-paris-a-new-coffee-mafia/

15 https://thebeansonfire.com/pages/about-us

16 https://melt-batignolles.fr/en

17 http://www.leshoublonneurs.com/

18 https://comme-chez-maman.com/

19 https://wim-a-table.com/

20 http://www.petitboutary.com/en/

21 https://www.ducasse-paris.com/en/professions/manufacture

References

Bellhoste, J. F., Lohr, É., and Smith, P. (2001), "Paris, ville d'usines au XIXeme siècle." In K. Bowie (ed.), *La Modernité avant Haussmann, Formes de l'espace urbain à Paris 1801–1853*, 338–52. Paris: Recherches.

Csergo, J. (2016), "The emergence of regional cuisines." In J. L. Flandrin and M. Montanari (eds.), *Food A Culinary History*, 500–15. New York: Columbia University Press.

Davis, M., and McBride, A. (2008), *The State of American Cuisine, A White Paper Issued by the James Beard Foundation*. New York: James Beard Foundation.

DeSoucey, M. (2010), "Gastronationalism: Food traditions and authenticity politics in the European Union," *American Sociological Review*, 75 (3): 432–55.

Escoffier, A. (1903), *Le Guide Culinaire*. Paris: French and European Publications.

Fischler, C. ([1993] 2016), "The McDonalization of culture." In J.-L. Louis Flandrin and Massimo Montanari (eds.), *Food: A Culinary History*, 530–47. New York: Columbia University Press.

Henault, S., and Mitchell, J. (2018), *A Bite-Sized History of France: Gastronomic Tales of Revolution, War and Enlightenment*. New York: The New Press.

Landaz, M. (2019), "Le « style Brooklyn » serait-ildevenuringard?" *Nouvelle Observateur*, Juin 30. Available online: https://www.nouvelobs.com//societe/20190 630.OBS15152/le-style-brooklyn-serait-il-devenu-ringard.html (accessed July 20, 2019).

MidiLibre. (2019), "Vingt ans après le 'démontage' du McDo de Millau: 'On avait la légitimité du territoire.'" Midi Libre, August 13, 2019. Available online: https://ww w.midilibre.fr/2019/08/13/vingt-ans-du-demontage-du-mcdo-de-millau-on-avait -la-legitimite-du-territoire,8358236.php (accessed September 3, 2020).

Parasecoli, F. (2019), *Food*, The MIT Essential Knowledge Series. Cambridge, MA: MIT Press.

Parkhurst Ferguson, P. (2004), *Accounting for Taste: The Triumph of French Cuisine*. Chicago: University of Chicago Press.

Parkhurst Ferguson, P. (2010), "Culinary nationalism," *Gastronomica*, 10 (1): 102–9.

Rouleau, B. (1985), *Villages et Faubourgs, de l'ancien Paris: Histoire d'un espace urbain*. Paris: Seuil.

Tomasik, T. (2007), "Gastronomy." In B. Kritzman, B. J. Reilly, and M. B. DeBevoise (eds), *The Columbia History of Twentieth-Century French Thought*, 238–42. New York: Columbia University Press.

Tramuta, L. (2017), *The New Paris, The People Place and Ideas Fuelinga Movement*. New York: Abrams.

Trubek, A. B. (2000), *Haute Cuisine, How the French Invented the Culinary Profession*. Philadelphia: University of Pennsylvania Press.

Trubek, A. B. (2008), *The Taste of Place, A Cultural Journey into Terroir*. Berkeley and Los Angeles: University of California Press.

Dispatch

London: A Typographic Stroll in Hackney

Adriana Rosati

Sunny days in London are rare but glorious. The whole city feels different and happier, and unassuming places turn radiant—even Iceland, the Hackney outlet of the drab frozen food chain. This store is next to Hackney Central station, and today is our meeting point. Maybe it is not a very good start for a stroll, as our goal is chatting about Global Brooklyn, but the sun can do miracles and Iceland vanishes behind the crowd of eccentric fashionistas and cyclists that in recent years have been repopulating the borough of Hackney in the East of London, for a long time a forgotten area of council estates, "greasy spoons," and kebab joints.

Both friends who are with me today live local: Itamar Ferrer, a service design director with Venezuelan roots, and Spike Spondike, a font developer and longtime East London aficionado. We share the experience of the Central Saint Martin's University of Arts (hence our interest for graphic design) and Spike is already in a nostalgic mood. "I remember taking the bus to go to Saint Martin's; I was living in upper Clapton at the time and this was the route of the bus."

We are walking down that very bus route, Mare Street, now a pedestrian area with lots of new places and cafés filled with MacBook-tapping young professionals. The look of the whole East of London has drastically changed. Once a rural outpost, the area suffered badly during the Second World War and its struggles reached a critical point in the 1970s and the 1980s when a reshuffle in the London economy wiped out most of the larger businesses, causing major unemployment and leaving only the lower-end enterprises and cheap warehouses. But in the last ten to fifteen years, things have been changing at a fast pace, due to the favorable position between Canary Wharf and the City—the financial hub of London—and a combination of then-affordable housing and workspaces, which enjoyed the active support of the Hackney Council to small London-based businesses and startups.

Spike loves this spot. "I must say that here a bit of a mix has been preserved, there are still places that were there fifteen years ago; the little pound-shop is still here, the McDonald's, the corner shop. It is a nice blend—I like it—I would have appreciated it if there was this atmosphere when I was living here."

On the other hand, the nearby Broadway Market has been completely taken over by cafés and restaurants. "Have you seen, Spike?"

"That was predictable," she adds, "but it's nothing like what happened to the old West Indian Brixton Market: it's just a food court now, no more market. No wonder there are some angry feelings between the local community and the new foodies."

Look how beautiful this shop front is, "Palm Vaults." Was it there before?

"Mmm . . . I don't remember it, but maybe they just rediscovered some amazing previous look and rebranded it. I really love this new appreciation for hand-painted signs. There is a trend in bringing back renaissance fonts and updating them, as well as distressed fonts that look engraved."

We are on Lower Clapton Road now, and I spot a place I have been reading about. "Look at this natural wine shop with a kitchen and a communal table: P Franco. They did something interesting here; they maintained the signage of the previous Chinese cash and carry and just placed their little logo on the glass door."

We all look up at the lurid yellow and red acrylic "Great Wall" sign in Cooper Black, the typeface of choice in the 1980s. Quite funny, their logo in contrast is very minimalist. How very casual!

Nearby, a group of tourists following a guide catches our attention; we wonder what they could find of interest here. Spike explains: "This was called the Murder Mile once, I think because there was a night club and lots of violent knife crimes happened in the past. Now the most dangerous thing you can find here is a barber! Oh, look, another one! They are everywhere!"

I suddenly picture my unfashionable neighborhood. You know what? It's just dawned on me that in areas that are not very cool and not revitalized—like mine—one every other shop is a women's hairdresser. In cool, gentrified areas, one every other shop seems to be a men's barber. How unfair is that!

"Good point!" Spike agrees. "In my area the first sign of gentrification was a cool barber, now there are three at short distance, just what we need!"

"This is Clapton Craft," Spike pushes us inside, "one of my favorite craft beer shops. I love the Bear Logo, I am a sucker for logos with animals."

Now, it's Itamar's turn to beam: "I LOVE craft beer labels, I love browsing them! They use lots of illustration, bold colors, and typefaces that look like screen-printed."

"Yes Itamar, craft beers have really redefined the way you think about packaging. So colorful, and I am really terrible because I love yellow and I'd buy anything in that color!"

"Tell me about it! I am the same, I'd buy anything with a green packaging; I think we, as graphic designers, are the most gullible clients!"

Recently, Itamar has been travelling a lot to Dubai for work and she is a bit skeptical. "I must say that these new places all look a bit the same. They use a very recognizable set of visuals, from the graphics to the interior decoration. I've noticed that even Dubai is full of cafés and restaurants that could actually be anywhere in the world. I am not sure I like all that sameness."

"Do you mean they haven't got a very strong identity?"

"Well, not that, but I think the world is getting smaller and more homogeneous because of social media; people are constantly communicating, everybody has access to lots of stuff, we can see clearly what is happening on the other side of the world, and this has a great influence. Also, certain ideas and a certain visual taste cross barriers because people travel a lot more. Travel is more accessible and easier than before so people see new things in different places; they may see something they want to import in the city where they live and prompt them to start new fads. Do you remember all that silly fuss around NYC 'cronuts' a few years ago in London? Everybody was going mad about them!"

I agree with Itamar, but I remark that access can also be a wonderful thing.

"Sure! Being a vegetarian and living in London, I noticed that here the whole plant-based movement has been influenced a lot by bloggers and peers that share experiences, information, and recipes; an army of small-scale influencers! This movement has a very specific image attached to it, social and health consciousness, support for the environment, and the visuals reflect that: very minimal and rustic, almost rough, because they align themselves with the idea of organic, which is often connected with gritty! They also favor recycled materials obviously, paper instead of plastic and a whole set of graphics that looks almost crafted or hand-painted or hand-written."

Indeed—I think aloud—the vegetarian and vegan ethos aligns itself perfectly with the recognition of craft and manual labor that is one important aspect of the Global Brooklyn's way of eating and drinking.

Itamar goes on. "One good thing is that this online circulation has helped a lot of small businesses. A British blogger I have been following for a long time called 'Deliciously Ella' started really small; she had some health issues and had to switch to a plant-based diet. Blogging was a way to deal with her health problems and reach out, and suddenly people started to follow her and ask about her recipes. If you look now, after only a few years, she's got a developed brand, three delis in London, books, a huge range of products, an app, and a podcast. Her brand image is really appealing. I liked it at first, because it's vibrant and colorful and knowing it comes from a personal struggle I was really attracted. But then I realized many other blogs look very similar . . . it's like they use the same template. I still like it, probably because I am their target audience, but I don't find it interesting anymore."

Spike interrupts us. "This alley is new too. Wow, I would have never imagined seeing a vegan café here. Shame they didn't space up the letters correctly in the logo! It's not very well done, is it? Maybe I can offer to respace the letters correctly in exchange for a couple of cupcakes!"

Oh dear, the kind of things that give headaches to graphic designers! This messy logo makes me think that the Global Brooklyn's graphic image is so well defined: it follows unequivocal rules and uses specific craft-style type fonts. Probably for this very reason anyone feels they can just get on Photoshop and design a logo. Like this badly spaced one. "An epidemic of amateurism!"

The same can happen with food. Simple food with amazing ingredients is the most difficult to make because it must be perfect. "I agree," Itamar adds, "this trend of craft, brush, script logos could allow some sloppiness. I think there are a lot of sketched-on-a-paper-napkin kind of improvised graphics. It's not always bad, but it can lead to an overall experience that's badly orchestrated. Maybe I notice it more because I am in service design."

I guess now a great care for the visual is a must, because food and its context always end up being photographed, shared, posted, reposted, and tagged. Consumers' activity on social media has become essential in the advertising process.

"So, being cool and instagrammable is part of the strategy of a place?"

"Definitely! On *Time Out* you find lists of 'the 10 most instagrammable cafés or restaurants in London' instead of 'the 10 best cafés or restaurants in London.'"

It's almost aperitivo time and we realize that—walking and chatting—we've ended up in front of Lardo, a Hackney restaurant near London Fields, the hearth of the Hackney revitalization. The restaurant is on the ground floor of a former industrial building with beautiful tall windows, lots of tables outside, and a beautiful hand-crafted wood-burning oven, covered in little mirrors, like a disco ball. Lardo's logo is very contemporary, bold and pink, and its walls are decorated with giant panels from up-and-coming artists.

We find executive chef Matt Cranston frowning in front of his Mac. "I was browsing for inspiration and always end up looking at the US scene, NYC's restaurants in particular, but lately I have been checking out Charleston. I love their pride for local traditions and the mix of old and new."

Any good ideas? You know, we are good guinea pigs!

Matt and I met a long time ago while working at game-changing restaurant St. John, the year it opened, and only recently we both were at its twenty-fifth anniversary party. Twenty-five years! Thinking about it, I feel I am having an epiphany! With its pared-down, postindustrial interior and its no-nonsense approach, St. John was a real precursor. The pride in the British products and specialties (until then mistreated), the refusal of any fussy food arrangements or decorations, the care in baking sourdough and making charcuterie, the search for exceptional suppliers, artisans, and craftsmen. These were all seeds of future ideas, many of which have now become part of the Global Brooklyn culinary paradigm.

St. John's chef and patron, Fergus Henderson, has spawned a big legacy that is now pumping life into lots of new places in London. To reword Mark Greif, Fergus wasn't an early adopter; he was the real avant-garde.

But the sun has set on Hackney and I can see an interesting plate of zucchini fritti coming our way. After all this talking, it's finally time to eat, drink, and be merry.

References

Clapton Craft: 97 Lower Clapton Road, London E5 0NP, UK.
Deliciously Ella: https://deliciouslyella.com/
Greif, M. (2016), *Against Everything*, New York: Verso Books.
Iceland, Hackney branch: 337 Mare Street, London E8 1HY.
Lardo: 197-201 Richmond Road, London E8 3NJ, UK.
P Franco: 107 Lower Clapton Road, London E5 0NP, UK.
Palm Vaults London: 411 Mare Street, London E8 1HY, UK.
St. John: 26 St John Street, London EC1M 4AY, UK.

5

Rio de Janeiro

Tropical Brooklyn, Global Botafogo

Thiago Gomide Nasser

This chapter begins with a disclaimer. For the past five years I have worked as one of the organizers of Junta Local, a platform based in Rio de Janeiro, Brazil, that brings together small artisanal producers, farmers, and businesses with aims of changing the local food system (Rossotto Ioris 2017). Our principal modus operandi consists of holding street markets—or "juntas," as they have come to be called—and events that mesh together the makings of a typical farmers' market and a street food party. There, one can be sure to find stands showcasing freshly picked *couve*, or kale (message to Brooklyn: it has been a thing here since way back), seasonal local produce, coffee beans roasted just yesterday and brewed by a purple-haired barista, and kimchi hotdogs. More than once or twice, I have overheard people at Junta pronounce the lines "this looks like that market I visited in . . . ." The blank can be filled in with Brooklyn, London, Berlin, Buenos Aires, or Tel Aviv. In our embrace of local food, we have been connected to others across the globe.

Ostensibly, not much separates us from other street markets in Rio. As far as infrastructure, we use the same makeshift wooden stands and line them up in public spaces, either well-shaded squares or streets. However, the atmosphere could not be more different. While the vendors that make up most street markets in Rio are in the majority resellers that bring their produce from CEASA (*Central Estadual de Abastecimento* or "State Supply Center"), home to the city's main conventional wholesalers, at Junta vendors are typically young folk who are pursuing a second career as artisans, food entrepreneurs, or cooks. Great care goes into researching products and design, reflecting upon the way stands are customized, menus are written, and packaging is developed. Vendors are keen to present the origin of their produce—usually local and organic—

and explain whatever expertise or personal experience shaped their creations. Concern with sustainability is a constant as well: styrofoam or plasticware are verboten and composting one's food waste is well advertised. Artisanal cheeses, craft beer, sourdough bread, kombucha, vegan butter, fruit leathers made out of *açaí*, tea, pastries, pestos, and charcuterie are commonplace features in our line-ups. In lieu of typical Brazilian street market fare in Brazil—the ever-present *pastel*, cheese-filled tapiocas—at Junta a passerby can stumble upon not only the aforementioned kimchi hotdog, but also falafel, ceviche, vegan burgers, and grilled cheese. The crowd consists mainly of millennials and young families who come to make sure their pantries and refrigerators will be stocked but might linger on for a light lunch and a glass of Brazilian wine, assembling around communal tables or picnic baskets on the sidewalk or a lawn. Our DJs are part and parcel of the experience as well. We have spawned a DJ collective that spins (and sells) vintage vinyl while they pump Brazilian grooves from the 1970s, Fela Kuti, and Steely Dan through the speakers. Although not enclosed by walls, Junta Local is proof that open-air spaces are also subject to deliberate design and convey meaning.

In tandem, we also organize an online platform and pickup system built to suit the needs of small producers, cutting the "middleman" and engaging the more conscious consumer. Meeting the producer in person at the market on a Sunday and buying through our website during the week is the kind of online/offline complementarity we aim for. This is worthwhile mentioning to the extent that buying directly from small producers and avoiding the pitfalls of the long-chain food system is a tenet central to the food movement (Finn 2017). The pickups occur during vacant hours in different venues around the city—a nightclub, a church, a former granary, and a collective workhouse. On set dates, usually Saturdays, producers that are part of the platform deliver their orders at early hours. A team that consists of volunteer producers checks and separates the orders before pickup time begins. Although the occupation of these spaces is rather functional, it provides another layer of proximity in the Junta Local experience. Consumers, by brushing against those who actually baked the bread or picked their organic arugula, and by watching the inner workings of this delivery system, are provided with a behind-the-scenes view that confers authenticity and concreteness to the lofty ideals we state on our website.

Putting together all these "platforms" with no past business experience and figuring out how to make ends meets without sacrificing values has led to the formation of our community—a 150-strong group of food entrepreneurs and producers that support our organization financially and otherwise and are called upon to weigh in on the rules and procedures that shape our activities. Our mission statement and the "manifestos" we periodically roll out on our website can be laden with the political calls to action that unite actors of the loosely knit but ever-growing food movement. Meanwhile, on Instagram we can wax poetic about seasonal persimmons grown on the outskirts of the city and feature a profile of a young couple that has retreated to the country to grow sugarcane and distill their own cachaça.

All of this surely sounds familiar to those that have been to Smorgasburg in Brooklyn, Markthalle Neun in Berlin, or shop at the local food coop or subscribe to a CSA box. In this regard, Junta Local is an instance of the cultural formation that the authors in this volume discuss as Global Brooklyn, connecting us to similar projects and businesses in Warsaw, Copenhagen, Mumbai, and elsewhere who wed together an ethical discourse on food and an aesthetic to go with it. As one of those in charge of this organization, I have perhaps been a witting and unwitting agent in the translation and adaptation of Global Brooklyn to the urban jungle that stands in contrast to the luscious forests and hills of Rio de Janeiro. At the foot of the iconic Sugar Loaf mountain with its Christ the Redeemer Statue is the neighborhood of Botafogo, where Junta Local started and most of our actions happen but also where there are other businesses, some of which we in some way incubated or incentivized. Although local beers and fare can be found aplenty here, it is also where, perhaps unbound by the clichés of samba and the beach, young urbanites, hipsters, millennials, and the like connect to the world and articulate ethical aspirations through the flavors of artisanal beer, kombucha, vegan milks, and locally made preserves. Many brands of these items that today are displayed in supermarket shelves got their head start at Junta Local. Also, one-time vendors at our stands have opened their own brick-and-mortar spaces and follow practices and discourses that echo Global Brooklyn. Interestingly, most of these new businesses and spaces have opened in Botafogo, a neighborhood that is located in the more affluent South Zone of the city but that differs from beachside, world-famous zones such as Copacabana or Ipanema. Despite their glow from a bygone era of bikini-clad girls passing by to the sound of Bossa Nova and the stream of tourists, Botafogo is where the hip can be found, sipping their flat whites and slurping their ramen as the drums of gentrification beat.

This report therefore benefits from close-up, first-hand experience but also may admittedly suffer from the lack of distance that might taint accounts drawn from participant observation. To create a sense of removal I will start taking a day off from Junta Local and assume the gaze and palate of a foodie-flanêur-ethnographer and begin analyzing other spaces and practices. These may or may not match the ideal type of Global Brooklyn, but will surely permit us to identify local adaptations and innovations that arise from the social, cultural, and economic specificities of the urban food scene in Rio de Janeiro. I will start looking at a specific product and space—bread at the The Slow Bakery—and then stroll and eat—my way through other spaces in Botafogo.

Global Brooklyn Bread at The Slow Bakery

It is 10:00 a.m. on a Saturday at The Slow Bakery. There is a long queue of people of all ages signing up for a table. The main sitting area is located in the back of the store. As they wait, most people settle on benches along the main entrance. On one side of the exposed brick wall, painted white, there are large black-and-white photograph prints of dough being handled by a muscular and tattooed forearm. Along the other side, customers can watch the actual mounds of bread being baked into crusty loaves in the humongous oven placed right at the front of the store. It is what the owners of this bakery located in a former repair shop, Rafael Britto and Ludmila Espindola, call "the factory." Once the loaves are baked, they are retrieved from the oven by way of a conveyor belt and placed on a tall metal rack with several shelves, like books in a library. This configuration not only signals the desire to blur the line between "front" and "back" of the house but to invert it altogether.

Clients who do not want to venture in for brunch can make their purchase and quickly head back to the street. Otherwise, once a spot opens, they can make their way inside. Before arriving at the main sitting area, they pass a "market" offering natural wines, artisanal cheeses, and other fare that can be taken home. Before sitting, brunch-goers still get to look through glass panels inside the "shaping area" located behind the ovens at the entrance. There, young apprentice bakers remove dough from the rising baskets, shape, and slash it before placing the loaves on trays and dispatching them to the baking area.

The sitting area consists of raw wood planks placed atop dark-colored metal frames. From the ceiling hang light bulbs wrapped with a malleable metal mesh. Although the trifecta of exposed brick, filament bulbs, and raw wood are there, the design does not feel contrived or cookie-cutter. Some tables are for two but there is also a large marble-top communal table. If the corridor-like entrance felt a little cramped, this area has a high ceiling with natural lighting flooding from the skylight at the top. Enough light pours in to keep alive a *jabuticaba* tree and also illuminates the horse-shoe shaped coffee bar, where baristas pour into cups. Along one of the sides there is a metal staircase. An arrow and the word "Academy" stenciled on the adjacent wall indicate what can be found on the second floor, a recently inaugurated "learning center" where the craft of bread is taught, together with other activities. Bread is not only the result of a knowledge-intensive process; the process itself, as part of the communal experience, is set to become a product.

Once seated, one can learn more about The Slow Bakery's mission before ordering. The menu is clasped onto a clipboard and the first page recounts its story and spells out its ethos. Although acknowledging that the inspiration for sourdough bread is imported, the local variety found here arises, literally, from the air of the local environment: "Our *lactobacillus cariocas* generate a complexity of flavors and aromas in the bread, but with much more determination, laid back attitude, and joy than their Californian sisters" (Slow Bakery 2019). When it comes to other ingredients used, the "farm to table" approach is also underlined: "Everything is fresh and seasonal, including the selected beans that are ground on order for the coffee we serve" (Slow Bakery 2019).

The acidity found in sourdough bread and its thick crusts is a departure from bread as it is usually consumed in Brazil. Although manioc and corn have a much longer history as staples in the national diet (Câmara Cascudo 2014), during the twentieth century bread made from wheat became the main source of calories for families. However, what is usually found in Brazilian bakeries is *pãofrancês* or "French bread," which started off as a version of a mini-baguette but as a result of massive industrialization and the use of flour premixes has become tasteless and cotton-like in texture. In this context, selling artisanally made, long-fermented breads in the beginning demanded not a small amount of proselytizing, which was achieved face-to-face at Junta Local and also through the savvy use of Instagram. In this regard, Rafa's and Ludmilla's original careers as an advertiser and script-writer, respectively, came in handy. Since the project began in 2014, the arc of their story has been conveyed through social media, alternating between close-up shots and carefully written narratives around their perils and feats. Rafa's business went bust and he decided to go for a drastic career change. He previously enjoyed carpentry as a hobby, and, despite never having baked before, was attracted to the manual aspect of bread-making. He started trying out recipes at home after stumbling upon a bakery in São Paulo that featured sourdough breads, a novelty in Brazil. He bought Elizabeth Prueitt and Chad Robertson's *Tartine* (2006) and intensified his research. His first loaves were sold online and in late 2014 he started participating in Junta Local, always selling out and establishing sourdough as a fetish for local foodies.

As the project flourished, the couple decided to open a shop, allowing Rafa to scale up his production and the Slow Bakery to materialize his ideals. In 2016 he and Ludmilla opened their first location on a street featuring several auto-repair shops in Botafogo. Rafa, assisted by his mother, an architect, and Ludmilla were personally in charge of designing the layout and the metal structures that were custom built to serve as displays for the bread. The space took on a postindustrial aesthetic: exposed brick, chalkboards, visible wiring tubes, a long concrete bench, raw wood tables, and an old beaten-down sofa thrown in the mix. Freshly baked bread could be taken home or one could stay for brunch: third-wave coffee from a local roaster, house-made yogurt, granola, and grilled cheese sandwiches made with *canastra* cheese. The foodies came in droves but also other well-to-do patrons started making the short trek to Botafogo as the store began getting showered by traditional media attention, marking a watershed in Rio's history (bread history, at least). Up to then, the selling point for the fashionable bakeries and cafés had been imported prosciutto or parmigiano reggiano. Bread, as long as it came in the form of a baguette or a ciabatta was an afterthought; coffee, as long as it had a fancy Italian-sounding brand name, was drunk black, strong, roasted to pitch-black to hide the defects of commodity beans.

At The Slow Bakery, locals were introduced to the several steps that make up the artisanal production of bread. The once anonymous baker who toiled in the back, now had an attractive face and a story to tell. Rather than elaborate and foreign-sounding menus, the offerings here are straightforward. You might order a simple grilled cheese or a salad, but the name of the producer and provenance of ingredients will be mentioned in detail. Despite the emphasis on locality, socializing or eating out for breakfast or brunch—peak hours at the The Slow Bakery—are decidedly an ascription to a globalized taste that has also taken hold in Brazil along the lines of social distinction. A couple years later, in 2019, The Slow Bakery moved to the new space described at the beginning of this section, allowing it to expand on its original concept, now in a setting in which other artisanal bakeries have opened in Rio and in other major cities. This constitutes a veritable "third wave" of bakeries, which is helping local foodies become fluent in *miches*, *fougasses*, *batârds* and the like, while transforming patterns of bread production and consumption. In this regard, bread-making is one of the manual forms of work that are being elevated to higher social status, as those with high cultural capital involve themselves in them (Ocejo 2017). However, learning the craft of bread has also been accelerated and democratized by the internet. Once dependent on long apprenticeships in highly regulated guilds or on expensive courses at a famous institute, now a youngster in Manila can watch YouTube videos by the high priest of bread, Michel Suas, post images of his loaves online, and discuss with a fellow dilettante in São Paulo what could have gone wrong with hydration. It is DIY bread culture writ virtual and global. In this light, The Slow Bakery's inauguration of the "Academy" might seem out of step. However, one of the distinctive traits that has been developed revolves around serving as a vehicle for diversity and social mobility. Ludmilla and Rafa have adopted labor practices that set them apart and expand their ethical discourse to include workers. In politically troubled Brazil, praising diversity on social media can be considered risky. In addition, many of those now employed by the bakery began as apprentices to Rafa and were selected from a bread-making technical course guided toward low-income students. The academy aims to amplify this capability while capitalizing on regular paying incomers.

In an emerging food culture that extolls the origin and locality of ingredients, there is a certain amount of irony in the fact that bread has played such a significant role: most wheat used by third-wave bakers in Brazil was cultivated far away, while one would expect that the same steadfast commitment to local ingredients would extend to the main component of bread. In this regard, the drive to meet Global Brooklyn standards can lead to several incongruencies, but also windows of opportunities, depending on your point of view or the timeframe being used. Most bakers avoid this contradiction by stating that the kind of structure required to make their beautiful bubbly breads can only be achieved using imported flours, whose composition might include wheat from Canada, Argentina, and France in order to match the technical requirements. Wheat in Brazil was a state-controlled commodity until the 1990s and small mills were killed off by industrial giants like Bunge and others, cutting local bakers off from the local supply chain (Calegar and Schuh 1988). Therefore, it would be preposterous to emulate the sort of partnerships between wheat growers and millers that many bakers in the United States, for example, have sought. However, bakers conscious of the foreignness of this archetype of bread, take different paths toward its adaptation or "localization." At The Slow Bakery, some kinds of bread that do not require high levels of hydration can be made with organic wheat grown in southern Brazil. Other bakers are more keen on using Brazilian wheat and have in fact launched campaigns to help growers plant it, an aspect that is well advertised. Yet others create breads that incorporate Brazilian ingredients such as *canastra* cheese, corn, manioc flour, and *pequi*.

Global Botafogo

However, bread is not the sole mainstay of Global Brooklyn in Rio de Janeiro. A whole world of different foods can be found a stone's throw away from The Slow Bakery. A few blocks east there is South Ferro, the brainchild of Sei Shiroma, a Chinese-Japanese Brooklyn expat who began his trajectory in Rio selling Neapolitan-style pizza pies from a mobile oven hitched to a car. After roving around the city for a while, Sei, who personifies Global Brooklyn, opened Ferro & Farinha, a pizza bar in the nearby Catete neighborhood. South Ferro is his second location, in Botafogo, with the more ambitious proposal of serving high-end fare fusing Asian and New York influences. Despite all the globalism encapsulated by Ferro & Farinha and South Ferro, the menus hew close to the gospel of local. One of his most famous creations is a pizza topped with crunchy *couve*, or kale, and soy sauce. The pies are no longer showcased at the Botafogo location as they were substituted with more Asian inflected dishes, with ramen and sushi prominently featured. The menu claims that the ramen noodles are made in-house and the fish is sourced from artisanal fishermen. The space occupied by South Ferro in Botafogo is minimalist; tapestries and Japanese iconography hang from the bare concrete walls. Pink-neon fluorescent tubes spelling "South Ferro" also hang from the wall, providing photo opportunities for Instagram posts.

Across the street and up Rua General Polidoro from South Ferro is Cru, a wine bar that specializes in natural wines. It opened in 2019 and the owners, Dominic Perry, a British linguist, and Selene Cruz, an architect, also run Winehouse, a more conventional wine bar, in Botafogo as well. Both fell in love with the "rawness" (*cru* in Portuguese means raw) of natural wines and their purity. As occasional participants at Junta Local markets, their stand customarily features a sign informing about the sole ingredient that goes into the wine they sell: "grape." Also at hand is a folder containing print-outs of the Brazilian legislation on wine. The listing of chemicals that may be added to wine (and not necessarily listed) for the sake of stability and conservation is overwhelming. Natural wines are technically unregulated in Brazil but a growing number of vintners are making the transition toward the bold flavors and funky aromas of fermented grapes. Cru is located in a renovated early twentieth-century house, many of which still exist in Botafogo. The outside façade was painted white and remains original but the insides were carved out to become a spacious bar, a dining room, and a mezzanine. Empty bottles, with playful labels that distinguish them from the solemnity of conventional wines, line the windows. The walls are exposed brick, tropical plants hang from the ceiling, and the wines by the glass are listed on a large chalkboard. Simple dishes made with local ingredients are served on ceramic plates locally made in Botafogo. There is no printed wine menu. The waiters (and not a sommelier) describe each available wine, sharing their own preferences and talking about the producers, many of whom Dominic and Selene personally visited or invited to tasting at Cru.

My next stop is Marchezinho, on Rua Voluntários da Pátria, Botafogo's main drag, that is lined by pharmacies, *botecos*, and by-the-weight restaurants. This bistro and bodega of sorts is partially owned and run by Sacha Mollaret, a French expat and architect who began the business in 2016. The atmosphere at Marchezinho is less minimalist than at South Ferro and Cru. The seating area is surrounded by shelf space with wine, jams, chocolates, and other products lined up for sale, as a reminder to patrons that this is not just a bistro but also a small market. Vinyl covers of forgotten Brazilian singer's records from the 1970s rests on a wall and flourishes of tropical plants convey nostalgia of a more gentle past in which singers of the *tropicalia* were at the forefront. Cupboards and other displays cases are made out of antique wood and add another dash of a past. Sacha designed most furniture himself but hired an experienced carpenter to execute. He also rummaged through antique shops in Rio's downtown to find items that allude to the style of old local general stores. The design contrasts with the postindustrial aesthetic of many Global Brooklyn spaces in the city and can be also found in new establishments around town and Brazil. In a much less industrialized country, it seems that designers and architects have mined other elements of the Brazilian past. During the day, the crowd that lunches at Marchezinho consists mostly of white-collar workers from the many office buildings nearby. At night however, backpackers, students, and couples mingle to sip on mojitos made with local rum and rosé wine from the southern state of Rio Grande do Sul, while enjoying menu items such as roasted palm hearts from a sustainable farm and local cheese plates. On your way out you can buy coffee beans from a small farm in São Paulo and sourdough bread from a local baker.

In a radius of a kilometer around The Slow Bakery the flanêur-foodie will also stumble upon a local market that eschews the use of plastic and sells local produce only (A Colheita), a pan-Asian restaurant that offers "authentic" adobo and pad thai (Chop-Chop), but also Michelin-starred restaurants and *botecos* selling cheap beer. That Botafogo has become the main culinary destination in Rio must be understood in light of the real estate market in the city and the search for cheap rent and space that is crucial to food businesses. Theoretically, Rio offers a myriad of neighborhoods outside the South Zone that could qualify as "food frontiers," as young chefs and fresh ideas look for an address. The port region and peripheral neighborhoods to the north of the city center are flush with old warehouses and out-of-use industrial spaces that are perfect candidates to the sort of renovation and gentrification that accompanies these businesses. However, limited public transportation and concerns with security discourage people and consequently restaurants, cafés, and bars from opening there. This scenario could have been reverted through public policy, and the revamping of areas and public transportation promised as Rio hosted a World Cup and an Olympiad in the span of two years, but all these projects failed. Botafogo is one of the few areas in the South Zone of Rio that was until recently impervious to real estate speculation, preserving the many layers of development that historically shaped it. In the late nineteenth century, aristocrats, wishing to distance themselves from the raucous and chaotic streets of the old downtown, built their villas in the then-distant Botafogo. In the early twentieth century, some factories were installed in the area, and housing for workers and managers resulted in charming complexes. As urban development expanded west, real estate in Botafogo stagnated, dictating that many of these houses would not be ravaged and replaced by high rises. Many of these houses are now home to spaces that are linked to their kindred around the world—Global Botafogo.

One of the original Global Botafogo places is Comuna, or the "commune," which moved to a house on Rua Sorocaba in 2011 as an attempt to find a place to host parties run by a group of college friends, while deep electronic grooves played in a lounge area where local artists hung their work and helped fabricate the chairs. In the past few years, Comuna has also hosted an indie bookshop, vintage cloth sales, and debates on transgenderism and DIY cosmetics. In the kitchen, burgers were prepared and sold to revelers from the start. However, they differed from those found at fast-food chains or late-night joints, which used industrial bread and meat patties. At Comuna, the bread was made onsite; the blends of meat used were carefully tested and sourced; ketchup and wasabi mayo were self-developed recipes, and the burger options ran the gamut from classical to one with sweet chili and bean sprouts. Through the years, Comuna's kitchen became more ambitious in terms of its sourcing and commitment to the use of local ingredients.

And here, personal histories once again intertwine. As a student and a Botafogo resident in the early 2010s, Comuna was a frequent destination for me. At the time I was researching the political elements in the discourse of Brazilian chefs and food media while working on projects to support small producers. I met Bruno, one of the owners of Comuna, together with his partner Tatiana, the kitchen leader. We bonded talking about the difficulty in accessing small producers and the lack of local ingredients even in high-end restaurants in Rio. A few weeks later, we were organizing secret dinners, a talk show about food, and also a small producers' market called Junta Local.

References

Calegar, G., and Schuh, G. (1988), *The Brazilian Wheat Policy: Its Costs, Benefits, and Effects on Food Consumption.* Washington, DC: International Food Policy Research Institute.

Câmara Cascudo, L. (2014), *História da Alimentação no Brasil.* São Paulo: Global.

Finn, M. (2017), *Discriminating Taste: How Class Anxiety Created the American Food Revolution.* New Brunswick: Rutgers University Press.

Ocejo, E. (2017), *Masters of Craft: Old Jobs in the New Urban Economy.* Princeton and Oxford: Princeton University Press.

Prueitt, E., and Robertson, C. (2006), *Tartine.* San Francisco: Chronicle Books.

RossottoIoris, A. A. (2017), *Agribusiness and the Neoliberal Food System in Brazil: Frontiers and Fissures of Agro-Neoliberalism* (Earthscan Food and Agriculture). London: Routledge.

Slow Bakery (2019), *The Slow Bakery: Panaderia Artesanal Carioca.* https://www.theslowbakery.com.br/

Dispatch

From Farm to Cup: The Emergence of Global Brooklyn Café Culture in Thailand

Yoshimi Osawa

From Bangkok to Chiang Mai

It was at a café restaurant, Man and the Figs in Sukhumvit, Bangkok, where Mr. Suradech shared his passion for coffee and clean food. The café serves artisanal drinks, including handcrafted kombucha, kimchi, and healthy muffins that he makes, as well as plant-based food cooked by his friend with whom he is opening a new shop. There was a print on a wall saying "Filter Coffee not People" by Department of Brewology, an American brand specialized in designing the artifacts of the specialty coffee world. Mr. Suradech, wearing a T-shirt of the same design, is a legal counsel who has recently set up his own business as a baker and brewer[1] in Bangkok. His passion for coffee and clean eating had developed not so long ago. It was when he was between jobs, which he described as a bad moment in his life, and he was in need of a space where he could be outside of his room. He began to spend time at Starbucks, and then his preference moved to smaller independent cafés for a better atmosphere. He claims to have learned the "original taste of coffee" at these places and has never gone back to Starbucks. His encounter with the world of specialty coffee has changed his life, giving him a new direction. Mr. Suradech reads "foreign" books (his own words) to learn about coffee and clean eating. He has also traveled to Kyoto, Japan, to visit specialty coffee shops and to Singapore to attend a coffee festival. Besides these influences on his ideas, knowledge, practice, and aesthetic taste, there was also Chiang Mai, a city in Northern Thailand where he took a coffee tasting class and established a new network of people, one of whom went on to become his regular roaster for the new shop.

Coffee in Northern Thailand

Thailand is one of the top coffee-growing countries in Southeast Asia after Vietnam and Indonesia, yet Thai coffee beans are consumed more domestically than exported. While Robusta is mainly grown in Southern Thailand, Arabica is grown in the north, where the specialty coffee scene is booming. Coffee was first introduced to Northern Thailand in 1973 by a joint program of the Thai Royal Government and the United Nations as a substitute crop for opium poppy cultivations to control the narcotics trade as well as to improve the lives of ethnic minorities living in the highlands (Pendergrast 2015). Chiang Mai, a central city in Northern Thailand, has been leading the specialty coffee scene, along with the capital Bangkok, both domestically and globally. Chiang Mai city is home to more than a hundred coffee shops, including a shop owned by World Latte Art Champions, who describe themselves as specialty coffee crafters. Another coffee shop is managed by families of the Akha ethnic minority, using beans from Akha villages. Their coffee was nominated for the World Cup Tasters Championship by the Specialty Coffee Association of Europe. These shops are frequented by a mixture of locals, expats, digital nomads, as well as domestic and international tourists.

Khwamsukh Café is a coffee shop that serves fresh farm-to-cup coffee as well as various types of hot and cold drinks and food, including the popular French toast. It is surrounded by trees and mountains in the Bo Kaeo neighborhood in Samoeng District, about three hours drive along winding mountain roads from Chiang Mai city. The café was opened about three years ago by a local family from the Karen ethnic minority and is now managed by a young niece of the family who brews, bakes, and cooks. The family grows coffee organically near the café and roast beans themselves. Due to the influence of the joint Royal-UN project, as well as Christian missionaries and international NGOs, poppy plants were replaced with coffee and other cash crops such as strawberries, which attract domestic tourists to the area.

The café has several Global Brooklyn aesthetic elements, such as concrete walls, black steel window frames, a blackboard menu, a brick counter, black pendant lightings, and rustic wooden and black metal furniture. While sipping an Americano along with an avocado smoothie on a Saturday afternoon, I saw smiling Thai customers taking photos with their phones at an Instagram-worthy section of the establishment featuring the logo of the café. I was told by a local who recommended the café to take photos at the same spot as it looked nice. There were also a group of bikers, local workers who left in the backs of pickup trucks, and older members of the local Karen community visiting the café. The atmosphere might give the impression of a hipster place on photos circulated on social media, but it is not one. The café is a popular location in the village, providing a public space that welcomes any types of people to casually gather. It is noticeable on their menu: their drinks and foods are considered reasonable; for instance, a latte costs 30 Thai Bahts (about US$1), which would be a fourth or a half compared to the prices for a similar coffee in Bangkok. This inclusive character is also visible in the selection of food and drinks, from foamed latte and colorful mille crêpes to simple Thai dishes such as fried rice.

From *Kafaeboran* to Global Brooklyn

Thailand's coffee drinking culture has gone through several phases, according to Professor Chayan Vaddhanaphuti of Chiang Mai University, who witnessed the cultural transformation himself. It can be traced back to classic *kafaeboran* (ancient coffee), made with dark roasted *oliang* coffee and served sweet with sugar and condensed milk. *Oliang* coffee is made not only from coffee beans but also from other ingredients such as roasted tamarind seeds, because it is simply cheaper this way. It was then followed by the introduction of instant coffee, which to a certain extent took place through the influence of American GIs, when US servicemen would visit Thailand during the Vietnam War. At the time, coffee shops were mainly social places for men. In the late 1990s, American coffee chains such as Starbucks and Au Bon Pain became popular and Thai chains such as Black Canyon and Café Amazon followed. Café Amazon is the largest coffee chain in Thailand owned by PPT company, a state-owned oil and gas conglomerate, with around 2,500 branches, many of which are located in gas stations. Thus, coffee drinking culture has spread across boundaries of social class, gender, urban-rural divides, and even ethnic differences in contemporary Thai society. However, different taste preferences do exist, along the lines of people's economic class and generation. While wealthier people drink coffee from chains and specialty coffee shops, average Thais like to drink sweet iced coffee from street stalls at a cheaper price and in a larger size (Pongsiri 2013: 2415).

Cafés in Bangkok and Chiang Mai show various aspects of the Global Brooklyn aesthetics that are outlined in the Introduction to this volume. In coffee shops in Thailand, I have observed certain features borrowed from global trends in terms of design and style, though they are limited to not only postindustrial fashions but also echo-rustic, minimalistic, and simple Muji or Scandinavian styles, with some occasional Thai or Asian accents, such as tropical plants and Japanese alphabet, *katakana*, being a part of the designs. There is no doubt that social media including Facebook, Instagram, and Line, a communication app commonly used in Thailand and other Asian countries, are some of the biggest actors in circulating images of eating and drinking that create transnational similarities.[2]

During my fieldwork in Thailand, I was told several stories of how social media has indeed impacted people's social and personal lives. To give some examples, a person in his twenties was debating between two persons to date because he liked one better, but the other had a greater number of Instagram followers. Another person was unfriended on Instagram by her friend because she did not "like" enough the photos that her friend posted. Having explored the Thai social media scene myself, I have noticed that not only does "camera eats first" behavior exist for social media purposes but people also tend to take photos of themselves posing at Instagram-worthy places. For example, some people may choose their clothes, hairstyles, and makeup based on visiting a certain instagrammable location such as a restaurant or café. Along with uploading photos of coffee shops or restaurants, it is popular in Thailand to use the check-in functions of social media to tell people where you are. There are numerous blogs and online review sites that tell which are good coffee shops to "check in." With or without having yourself featured in a photo, aspects of life related to eating and drinking are seemingly considered to be worth sharing with others, as if one could be defined by where, what, and with whom one eats and drinks.

Other aspects of Global Brooklyn might present differently in Thailand. What makes Thailand, particularly Chiang Mai, unique compared to other places is that the coffee is grown in the area, which makes farmers, roasters, and baristas physically close to each other. There are recent cases of baristas or coffee shop owners growing their own coffee beans. This will bring the idea of revaluation of labor that has been observed in other Global Brooklyn-style locations to a different level. The owner of a coffee shop in Chiang Mai, for example, emphasizes that he can have better control over the quality of beans by growing his own. What we observe here is that there is trust in manual processes, particularly in drawing on embodied practice and sensory knowledge.

There are public entities that have officially been involved in the transmission of knowledge and skills related to various aspects of coffee from farm to cup in Chiang Mai. Lanna Thai Coffee Hub was launched in 2018, initiated by Chiang Mai University with governmental aid. The Hub aims to provide information, knowledge, and support to local farmers or anyone interested in coffee, in order to increase the quality standards of the crop in the region. The Hub provides assistance to entrepreneurs through branding and marketing, or even barista training.[3] Coffee has also been seen as a tourist attraction, as travelers visit coffee shops, and in some cases even plantations and roasters. Thus, behind the rise of specialty coffee in Chiang Mai, there is also the perception that the product constitutes an important source of income for the region and the country, with high potential for further growth due to an increasing demand for specialty coffee globally. There is also generally a positive attitude toward entrepreneurship in Thai society, and family entrepreneurs are viewed as having particularly strong positions in the country's economy. Thailand is also known for its culture of street food and drinks, which is mostly based on small-scale ventures. These aspects of Thai society in relation to the economy, business, and food and drinking culture might have encouraged the opening of a large number of specialty coffee places in Thailand. Coffee farmers can in fact earn more by roasting, and even brewing and serving coffee themselves, than just by selling cherries and parchment. A family member of a local farmer in Chiang Mai who grows coffee told me that if they can roast their coffee themselves, they can earn more by selling it at a higher price. However, her family does not because of the investment necessary to set up the roasting facility. What we can see here is a challenge for farmers to deviate from the pattern of remaining simple laborers by acquiring knowledge and skills to grow and produce higher quality coffee beans and by meeting certain standards. Thus, the emergence of the Global Brooklyn coffee culture has impacted the local economy in the coffee-growing area.

Another aspect of the Global Brooklyn regime, the shift in taste judgment discussed in this book's Introduction, needs a careful analysis in Thai cases, mainly due to differences between Bangkok and Chiang Mai that reflect internal social, economic, and historical dynamics. While people in Bangkok tend to have urban lifestyles and cook less at home, benefiting from their vibrant street food culture, in Chiang Mai, particularly in suburban areas, people more commonly keep homecooking traditions. It is common in Northern Thailand that people cook food using charcoal and firewood and get most vegetables and herbs from their own gardens. However, I have heard both in Bangkok and in Chiang Mai that people emphasize the better taste of food when cooked with "real fire" rather than gas or when prepared with a stone mortar rather than a blender. At the same time, older generations show their concern that the youth might not understand these differences anymore because of their westernized and globalized diets and taste, as well as their larger consumption of mass-produced food.

Global Brooklyn is indeed mostly produced by younger generations. There has been a recent shift in preferences, as seen in Mr. Suradech's case, with the development of new taste based on valuing artisan craft coffee more than chain coffee. However, some might be keeping a preference for enjoying the flavor of charcoal cooked meals as well as locally sourced fresh coffee within the same sensory arena as in Chiang Mai's case. Thailand has gone through the industrialization process at a more rapid pace compared to the Global North (Komin 1995), which has resulted in creating mixed social structures within the country. Though Thailand has been categorized as a Newly Industrialized Country as a whole, the social structures are different between Bangkok, mostly a postindustrial urban area still surrounded by industrial and agricultural communities, and Chiang Mai, a mixture of agricultural, industrial, and postindustrial characteristics in one place. Even just within Thailand, the meaning of the Global Brooklyn phenomenon reflects different elements of society, which brings an opportunity for us to consider how complex our eating and drinking practices have become.

Notes

1   Mr. Suradech chose the term "brewer" rather than "barista" as he brews not only coffee but also other types of drinks such as kombucha, beet kvass, and kefir.
2   Social media use is considerable in contemporary Thai society. According to recent data, almost 80 percent of Thai population are actively engaged in using Facebook, and the number of Instagram users takes seventeenth place in the world (Tangsuwan 2019).
3   Lanna Thai Coffee Hub, https://www.lannathaicoffeehub.agri.cmu.ac.th.

References

Komin, S. (1995), "Changes in social values in the Thai society and economy: A post-industrialization scenario." In M. Krongkaew (ed.), *Thailand's Industrialization and Its Consequences: Studies in the Economies of East and South-East Asia.* London: Palgrave Macmillan.

Pendergrast, M. (2015), *Beyond Fair Trade: How One Small Coffee Company Helped Transform a Hillside Village in Thailand.* Vancouver/Berkley: Greystone Books.

Pongsiri, K. (2013), "Market feasibility for new brand coffee house: The case study of Thailand," *International Journal of Social, Behavioral, Educational, Economic, Business and Industrial Engineering,* 7 (8): 2414–17.

Tangsuwan, K. (2019), "Thailand Zocial Awards." Available online: https://thailand.zoc ialawards.com/2019/keynotes/

6

Constructing New Communities

Global Brooklyn in Tel Aviv

Liora Gvion

Introduction

On a commercial street of Tel Aviv, known for its inexpensive stores and eateries, Uriah operates a café that is recognized by its customers as an emblem of the local Global Brooklyn scenario. One enters the place via a dark hallway of a falling-apart building, which leads to a backyard full of old-looking wooden tables, large enough to fit six to ten people, chairs, and benches. Meni's place, a fifteen-minute walk from there, operates in what used to be a poor and unsafe neighborhood, now attracting youngsters looking for affordable housing and entrepreneurs seeking business opportunities, alongside an art gallery, bars, a bookstore, and several small bistros. Unlike Uriah's café, Meni's place is a hyper-designed space that features high ceilings, bare walls revealing the pipes running along them, bins containing various types of coffee beans, and small tables made of old wine barrels. Behind a wooden counter, in both settings, a certified barista who looks like a hipster prepares what for him is the ultimate cup of coffee, while customers specify the type of beans, roast, and milk, and drink before they sit wherever there is a free seat.

Although attuned to global trends, Global Brooklyn Tel Aviv takes a different direction as it exemplifies the decline in the Israeli middle class, the complicated position of the Israeli millennial generation, and their quest for social belonging. The local notion of the middle class implies ownership of property, a steady professional job that offers a stable income, benefits and often tenure, and the means for enjoying leisure activities. The shift from a welfare state to neoliberal politics has created a precarious job market with little security; fewer returns on academic credentials, which cause nestling in marginal or temporary jobs; and a rise in the costs of real estate and health services—all of which implies that their aspirations to live a middle-class lifestyle have been frustrated. Consequently, many Israeli millennials are facing increasing socioeconomic insecurity (Mohar 2015). As are millennials around the world, they are unattached to organized politics, less engaged in civic and political participation, less patriotic than the older generation, critical of neoliberal capitalism, and sensitive to the discourse of intersectionality (Milkman 2017; Taylor et al. 2014). Global Brooklyn Tel Aviv enables them to widen the social distance between themselves and the original inhabitants, allowing them to "act middle class" while frequenting consumption spaces that require modest expenses but reposition them within the symbolic boundaries of the Israeli middle class.

For the owners, most of whom are middle-class businessmen in their forties, Global Brooklyn Tel Aviv is a business opportunity to target millennials living in gentrified neighborhoods and seeking spaces to socialize with peers. There, as pointed by Halawa and Parasecoli (2019) employees and customers acquire and articulate a shared language, manifest their cultural capital, and develop social relations that evolve into concrete communities and cultural affinities. Although not beneficiaries of the global economy as their parents were, it is in these establishments that millennials, who share expectations and values that reveal new forms of cosmopolitanism, participate in a global culture and engage in activities similar to those of young people in the Global North.

Most places reflect a preference for wood and a nostalgia for a less virtual world. Rather than displaying a heavily postindustrial design, these spaces are often planned to connote authenticity by mobilizing historical images of the gentrified neighborhood. The original infrastructure, in combination with old-looking wood, objects, and furniture, often collected from the streets nearby, is associated with that image. Digital media is used for communicating and disseminating information and food knowledge and for reaffirming the customer's identity. Although visitors are active users of social media, owners cannot rely on online communication and virtual communities for publicity as their customers mostly learn about the establishment while walking the streets looking for locales to use as extensions of their homes, a work space, or places to socialize.

The sensual experience is often traded for serving good yet unsophisticated food and beverages in a "correct" atmosphere to cater to millennials who eat out on a daily basis and to occasional visitors, assuring that everyone finds something to eat. Rather than promoting local and traditional specialities, owners see "local" as a commitment to buy merchandise in stores owned by the area's original inhabitants. Finally, as elsewhere, Global Brooklyn Tel Aviv rests on an ethos of manual labor, previously discounted as a working-class activity, but now alluring and newly appreciated by urban creative middle classes (Halawa and Parasecoli 2019). By granting cultural capital and appreciation to these occupations, Global Brooklyn Tel Aviv enables millennials to settle into jobs to which they attach lifestyle practices and symbolic capital.

This chapter describes Global Brooklyn Tel Aviv from the point of view of fifteen owners, whose establishments conform to the abovementioned features, and eight baristas, chefs, and bartenders. All were chosen after discussing Global Brooklyn with my students and asking them to identify places in Tel Aviv that resonate with the concept. Some places, by the time I reached them, had closed or became different settings. I interviewed individuals in the remaining establishments. The interviews, which lasted two hours on average, were scheduled after I had spent a couple of hours in each place and taken notes. I talked with owners about their reasons for opening the place, the targeted clientele, the role their place played in their lives, and their considerations when designing the space and composing the menu. I talked with employees about their reasons for engaging in working-class occupations and the knowledge embedded in their craft.

Gentrification, Global Brooklyn, and Tel Aviv

Gentrification is a process in which neighborhoods populated by marginal groups witness an influx of young educated individuals of middle-class background who are looking for an affordable urban lifestyle. Places are reimagined as the creative centers of a symbolic economy, and business opportunities accommodate the gentrifiers' notion of authenticity, granting them consolation for having little chance to gain the rewards of powerful elites (Hubbard 2016; Zukin 2008). Newcomers—owners and employees—usually work in the creative industry or cafés, bars, and music, food, and fashion stores. They deploy their cultural competence and taste to redefine petit bourgeois occupations as expressive of cultural capital and status rewards (Schiermer 2014; Scott 2017) and refigure local taste cultures (Cronin, McCarthy, and Collins 2014; Hubbard 2016; le Grand 2018; Maly and Varis2016; Michael 2015; Schiermer 2014; Scott 2017; Stahl 2010; Tuz 2014–15). Having lost their economic capital and the advantage of the middle class, but possessing educational and cultural capital, they navigate diverse cultural codes, seeking to reassert their individual and social identities through the acquisition of culinary and cultural capitals (Greif 2010; Hubbard 2016; LeBesco and Naccarato 2015).

Neighborhoods such as Florentine, the old central bus station area, and Shapiro, where mostly Mizrachi Jews who were poor manual laborers lived and worked, have witnessed the departure of the successful among them. This was followed by the arrival of African migrant laborers, asylum seekers, and political refugees, who were seen by the marginal populations of the area as contributing to the neighborhood's deterioration. Gentrification has been slow and is seen both as a problem, raising the cost of living and changing the social makeup of the community, and as a means of rescuing the area. In Jaffa, an Arab area in which some Bulgarian and Mizrachi Jews settled in the late 1940s and the early 1950s, gentrification has been more complicated. The departure of affluent Jews, due to a constant neglect of the area by the Tel Aviv municipality, has contributed to an increase in crime rates and Jewish-Arab tensions. With the arrival of the gentrifiers, property values have gone up, new consumption spaces have opened, and Arabs have been feeling pushed out by affluent Jews seeking luxurious accommodations and by middle-class individuals and families seeking affordable housing.

Sites of leisure, such as Global Brooklyn Tel Aviv, serve as safe zones for newcomers to meet and distinguish themselves from the older residents of the neighborhoods, with whom they have little contact. However, operating in a small yet competitive market requires entrepreneurs to adjust the spaces to their customers' needs. For instance, they all design the space to form specific communities. These are essential because many of the regulars are new to the city and wish to meet people and make friends. They use these spaces to engage in an active social life with people like themselves and as platforms for promoting social change.

Designing a Business and Constructing a Community

Converting old buildings or workshops on the narrow streets of old working-class neighborhoods to accommodate newcomers pushes the gentrifiers to design spaces that new inhabitants see as reflecting an authentic image of the neighborhood. Capital is used to make the place look as if little was invested: old-looking wood, original tiles and floors, old furniture, unpainted walls, and old objects removed from their original contexts, such as a trumpet turned into a tap or a washing machine converted into a large vase. Menus are written on a chalkboard, and images of the dishes are shared by customers, chefs, and owners via social media.

Although targeting digital natives and occasional weekend tourists, the design communicates messages about using the place to generate face-to-face interactions and form tangible communities. For example, entrepreneurs often accept registered letters or parcels for customers while they are at work or store an extra set of keys. In order to prevent hipsters or parents with babies from frequenting their place, they do not install electrical outlets that would enable visitors to charge their devices, and they make sure there is no room for strollers or high chairs. Joseph, who operates such a place in Jaffa, says:

> It used to be a glass workshop and I wanted people to get the feel of it. I invested a lot of money so it looks old. I don't want hipsters who would use it as their workplace, ordering nothing but a cup of coffee, or babies who make noise and run around. I want people who appreciate the design and what we serve.

To open a high-end bistro, Jaco broke the walls between three adjacent old stores, paved the floor with concrete and filled the space with furniture he gathered from people who left the area. He says: "I only did what was necessary to accommodate the space to my needs. I encourage customers to draw or write on the walls and let painters hang their paintings. I want customers who appreciate the design and the dishes we serve."

The design, then, is meant to articulate the locale with the material and social environment. While Joseph and Jaco both want their places to look "old," yet reveal the physical and financial investment, they each use the design to attract a diverse clientele. Joseph communicates his preferences for wealthy gentrifiers and tourists, whereas Jaco targets youngsters, members of the creative classes whose habitus enables them to appreciate the space and the dishes he serves.

In addition to communicating a message about the culture and social makeup of the place so that newcomers recognize it as "their own," the design enables the formation of a community through which inhabitants promote further changes. The more customers think of the place as enabling social connections and a social life, owners claim, the more they frequent it and work together to change the neighborhood. Uriah relates: "I don't put sugar or salt on the tables or hang signs directing to the restrooms, so people will talk to each other. We don't have tables for two, so people sit together and interact. A lot of demonstrations were planned here."

Operating from a shed by a large yard full of old furniture, Michael generates interactions among customers by deliberately assigning only one staff member to take orders and making people wait for an hour for their drinks. He says: "It's a social café, a place where people from the area come to spend time. They don't have to order anything. We have movie nights and poetry nights, yoga classes and a communal garden where residents grow vegetables and take whatever they need."

Omer, who operates an upscale pizzeria, lets his customers rearrange the space according to their needs: "People move the furniture around while they're here. We encourage them to have book parties or exhibits here. We put a refrigerator on the street with the food we can no longer serve so the poor and homeless can survive on our food. This is what I call social responsibility."

In encouraging social interactions, communal activities, and ongoing changes to accommodate customers' needs, owners apply a communal managerial philosophy, which transcends the traditional owner-customer relationship. Customers are invited to use the place as an extension of the home and therefore take responsibility for their social and physical environment. Uriah, for example, attended to the suggestions of his customers, most of whom have traveled the world, to incorporate what are known as global or hipster items: "The customers suggested I bring kombucha, almond milk cocoa, matcha, and jump, a lemon ginger drink seasoned with honey and hot pepper."

While this design is attractive to newcomers, the owners acknowledge the alienating effect it tends to have on the original residents, who mostly do not fit in. Daphna talks about the difficulties of operating a place that can attract young educated urbanites as well as socially marginalized populations: "We have those who moved in because it's affordable, construction workers, junkies, alcoholics, poor people and prostitutes. A place should fit the environment. So I have formica tables, old lamp bulbs, bare walls and high ceilings."

To conclude, although the owners plan their places so they articulate with the social and physical environment and connote authenticity, they admit that the design is also meant to communicate messages about their targeted clientele. Aware of their customers' modest financial means and active urban lifestyle, they design their place to constitute a community and construct a social café that functions as a safe zone for the new residents by enabling them to act middle class and serve as agents of social change, even while they have temporarily lost their economic capital and the background dominance of the middle class.

Showing Expertise

Global Brooklyn Tel Aviv provides a sensory regime that expresses a renewed relationship with locality. The appreciation of original and local taste, stemming from the possession of certain knowledge and cultural capital, enables consumers to understand and enjoy the experience, and turns into a means by which they distinguish themselves from other social groups. Through affordable consumption that takes place in these establishments, identities are constructed, negotiated, and expressed.

Given the limitations of the market, owners in Global Brooklyn Tel Aviv adjust the foods and beverages to accommodate customers' tastes, expectations, and financial means. Owners, who get their updates from social media and meetings with growers, distributors, and entrepreneurs from around the world, have learned to appreciate acidic coffee, oxidation, and strong smells in wine, fruity beers, and sophisticated dishes. However, realizing that catering only to high-end taste preferences is not necessarily profitable, they take it upon themselves to gradually educate consumers while continuing to offer varieties more in keeping with local tastes.

Take the case of coffee. Owners and baristas often attempt to teach customers to appreciate new tastes in coffee. Yet, it is their recognition of the local preference for bitter coffee and its consumption with baked goods or a sandwich that ultimately guides their decisions. Ran tells us: "We have six different types of coffee that we roast to various degrees. It doesn't make sense to have only one kind of coffee and a selection of teas and wines. There's a whole world of coffee to which I can expose customers, but many of them want nothing but a tall cappuccino or a bitter espresso."

Meni's clientele has a similar attitude: "Most people don't understand there's a right taste to coffee and they should develop a preference for it. One should match his coffee to the time of day and the type of weather. Many Israelis don't know that bitter coffee is over-roasted coffee. It'll take time until customers learn that good coffee is fruity."

Although owners and baristas are up to date with recent perceptions of good coffee, their customers, they argue, are lagging behind and have developed neither the taste nor the appreciation for coffee as people drink it in other global cities. Abner says customers also fail to understand that different regions produce different tastes in coffee: "Italian coffee is bitter and African coffee is fruity. I use different beans and roasts for different drinks. If you use Ethiopian coffee to make espresso, it'll taste sour. When drinking French press or drip coffee, you're better off with Ethiopian beans. What some call acidic coffee, connoisseurs recognize as fruity."

Taste in Global Brooklyn Tel Aviv, then, is an acquired trait manifested through a language that reveals how connoisseurship is negotiated with the local palate. The latter stems from an interactional process in which customers are exposed to different types of beans, drinks, vocabularies, and habits, which enable them to reinterpret certain palates as "right" and testify to their possession of a certain cultural capital that is locally specific.

Preparing the "right" kind of coffee requires owners to follow certain principles when choosing the beans, growers, and brokers. Although wishing they could follow the principles of ethical and fair trade, thus being attentive to another global trend, catering to a small market makes it unprofitable. Abner settles, therefore, for buying either from growers or from honest mediators: "The owner travels around the world to find growers. We work with farmers in Tanzania, Ethiopia, and Guatemala who grow coffee especially for us. We engage in direct trade to make it less expensive for our customers. We change the beans every season. In the spring we serve Guatemalan coffee."

In addition to locality and seasonality, as an identifiable feature of Global Brooklyn Tel Aviv, Ran considers the way in which the workers are treated: "When I visited Ethiopia, I met an English broker who lives and works with the local farmers. He treats his workers as his partners. I, too, am now involved in fundraising for the village. For me this is much more than fair trade."

As in other global cities, locality and taste are distinctive features, which emerge out of social relationships embedded in choosing the coffee origin, marketing, and producing coffee. But locality in Global Brooklyn Tel Aviv also takes a different meaning: buying ingredients from stores owned by the old residents of the area. Aside from being convenient, buying small quantities and thus reducing the amount of food thrown out every day, this decision presents entrepreneurs as concerned with the neighborhood and wishing to be part of the local economy. Amos relates: "We serve sandwiches, quiches and salads. All the ingredients are bought from the stores on our street. The sourdough breads, cakes, and baked goods are from a bakery a couple of blocks away."

For Michael, shopping locally is part of an overall ideology that calls for collaboration, as all residents, new and old, are part of the same economy and consequently have shared interests:

> I was determined to make sure the old residents don't go out of business. It's a win-win situation. Unlike the past, when the old residents were pushed out of the area, I want them to stay and benefit from our presence. It's trendy now to eat and drink locally and provide customers with information about the origins of their food. I can serve food that is related to the place where it's consumed.

Locality, then, translates into storytelling by revealing the social relationships embedded in producing the foods served. The story, in Michael's opinion, has two facets. It is a story about social responsibility that gentrifiers have for the original residents and a story about changes he wishes to see in the dynamics of gentrification, interwoven into neoliberal politics as the privileged (rather than state agencies) see themselves as accountable for the well-being of the unprivileged. Following the same principle, Daphna chose to work with a Palestinian roaster from the occupied territories: "I was determined to buy supplies from people who can easily lose their business. I've been buying for years from a Palestinian who was trained in Italy, bought a franchise to open a place in his village and provides me with great coffee."

In buying coffee from a Palestinian roaster, Daphna sees her business as enabling her to expand the meaning of locality and the boundaries of social responsibility to encompass members of marginal social groups. Mutual dependency and trust not only enable business transactions but also allow skilled and reliable people from marginal populations, such as Palestinians and old Mizrachi Jews, to partake in the social collective as individuals, rather than as representatives of distinctive communities.

The discourse on coffee somewhat resembles the talk about high-end hummus, made of high-quality garbanzo beans, whole-sesame tahini, real lemon juice, and prepared by experts who have mastered the skills for making a tasty product. Saul says: "Good hummus is made of the best ingredients in the market and is served fresh. We precook the beans, but every thirty minutes our cook prepares fresh hummus, using a mortar. The same applies to the chopped salad and the coffee."

Storytelling is used to reveal the amount of knowledge, apprenticeship, and devotion embedded in making good hummus. Tomer shows his customers a cartoon about a hummus bean that spent its days in his father's hummus place. On his deathbed (drawn as a big pot of cooked chickpeas), the father, a big chickpea, tells his son it is time he too turns into good hummus.

As much as owners and employers value expertise and high-end dishes, in catering to millennials with limited financial means, but of middle-class background and with high cultural capital, they also have to consider their menus carefully. Most of them, as already mentioned, settle for serving unsophisticated, filling, and moderately priced dishes in a place to hang out. Matan and Ori, the owners of a high-end restaurant, apply a different strategy: "We have a pay-as-you-please policy. Most of our diners appreciate the dishes and don't take advantage of our policy. There are enough people in Tel Aviv who recognize sophisticated food and pay for it. We tolerate the few who do not."

Matan and Ori, then, attribute knowledge and social capital not only to the chef but to the clientele as well, which is able to recognize the quality of food and translate it to proper monetary value. Customers who do not hold the necessary culinary capital to enjoy the food usually feel out of place.

Appreciation for the food and beverages served in an establishment is manifested by customers as they co-produce value by photographing the locale and the dishes, tagging their friends, and sharing the images with peers on the internet. Most owners write up their menu on a chalkboard and post it online, knowing they can rely on their customers' comments and photographs for publicity. Jaco relates: "Every day, after the chef selects the ingredients and decides on the menu, I post it online. Sometimes I add a picture of a dish. Some guests arrive knowing what they want to eat. They take pictures of the food and post them online immediately. There are new pictures every day."

Global Brooklyn Tel Aviv, then, is the foodway of millennial digital natives who, in cooperation with owners and employers, transmit, adapt, and make use of physical and digital spaces. Food and beverage knowledge are jointly produced, disseminated, and reinforced through expertise, ethical consumption patterns, and a conversation about food and lifestyle that enables the sustainability of locales. These adaptations make it possible for Israeli millennials to become part of a coherent aesthetic and provide them with an opportunity to live a global urban lifestyle while also supporting the original community. At the same time, Global Brooklyn Tel Aviv provides millennials with the infrastructure to surround themselves with people engaged in constructing an identity that distinguishes them from the original inhabitants of the area.

The Reevaluation of Working-Class Occupations

Global Brooklyn Tel Aviv is further constructed and reproduced via the work of millennials, most of whom are of middle-class background or from the creative class, who engage in and come to appreciate what used to be considered working-class occupations. By reevaluating working-class occupations, millennials claim cultural capital (Bourdieu 1984, 1986) that guides their actions and tastes, objectifies cultural goods, and underscores social differences.

Both owners and employers claim millennials are realistic about their prospects and aim at living a modest yet up-to-date global and urban lifestyle. Rather than seeking upward mobility, they interpret their social position as generating an agency call to advocate for social change. The change revolves around inserting alternative meanings into work, valuing craftsmanship, and interpreting a modest lifestyle as enabling engagement in leisure activities. My interviewees spoke about the knowledge, precision, language, and value embedded in manual labor and how it gains the appreciation of their customers. Shai, a barista, tells about the respect his knowledge wins him: "I have a customer who comes in twice a day. The first time she came, she watched the staff for about thirty minutes. When she finally ordered, she insisted I make her coffee, claiming she could tell by my movements that I was a qualified barista. It was a great compliment."

The acquisition of knowledge, recognized by bodily movements, Ran says, wins customers' respect, teaches young people the meaning of a vocation and guarantees a certain income: "Having worked in the coffee business for years, I've come to appreciate high-skilled craftsmen. There's a lot of knowledge in their hands, which they have accumulated in years of working. I respect them more than I respect people who develop new applications. These people will never go hungry."

The redefinition of working-class occupations, as enabling financial security and embedding cultural capital and embodied knowledge gained through apprenticeship, makes it possible for millennials to work in the food and drink industry. However, they work mostly in locales that resonate with Global Brooklyn Tel Aviv. Gil says:

> When I tell people that I'm a cook, they say "no! You're a chef." The truth is I'm a cook. I knead dough, I clean fish, I carve meat, and I sweat by the oven. My customers often come to watch me work and comment, with appreciation, about the way I move my hands and body when I cook. They see my work as different from the work of the other pizzerias in the area.

For one to be recognized as a qualified worker in Global Brooklyn Tel Aviv requires more than mastering manual skills, knowledge, bodily movements, and gestures. All of these must be recognized by connoisseurs as work embedded within the necessary symbolic capital that distinguishes Global Brooklyn Tel Aviv from other kitchens and wins appreciation and prestige.

By being qualified to work in Global Brooklyn Tel Aviv and willing to live on moderate means, employees gain another benefit: free time to engage in leisure activities. Nathan tells us:

> I work six or seven hours a day, which gives me time to go to the beach every day, play the piano, read books, and spend time with friends. I'm happy with my life as it is, knowing I can always find a job as a barista. When I know all there is to know about coffee, I may open my own place.

According to Omer: "We believe that since we're unlikely to find a good job with benefits, we are free to live our lives the way we want and be content with the little we have. We work to pay our rent and bills and to be able to enjoy the free time we have."

Workers in Global Brooklyn Tel Aviv, then, are not book learners. They learn, process, and verify their knowledge through apprenticeship and ongoing manual work that reveals itself through bodily movements, "correct" vocabulary, and rules regarding matters of preparation, all of which win them prestige and respect from connoisseurs. They become part of a community of workers and consumers who share a language and uphold similar cultural capital. Once manual activities are redefined as expertise which few can acquire, youngsters feel positive about engaging in occupations previous generations of similar backgrounds would not have considered. This position further generates an agency call to promote social change that redefines the social value of work and suggests options for living a fulfilling life.

Conclusion

Cultural globalization in Israel, that is, the entrance and adaptation of global trends into daily practices, was originally promoted by middle-class professionals and the creative classes, who supported the import of global practices into local cultural spaces alongside an ideology that advances consumer culture and a bourgeois-bohemian global lifestyle. This chapter has laid out the distinctive features of Global Brooklyn Tel Aviv and the ways in which the latter provides Israeli millennials with the opportunity to partake in a global culture to which local meanings have been attached. For such millennials a global culture, in a localized version, is part of their habitus. Operating in a precarious and unstable labor market, they find that Global Brooklyn Tel Aviv not only constructs and reaffirms an identity that encompasses global and local features but also provides compensation for the loss of a middle-class lifestyle attributed to changes in the economy, followed by fewer returns for academic credentials. Their position is facilitated by their privileged status in comparison to other social divisions.

Entrepreneurs cater to this small market of millennials, who possess the "right" cultural capital but have moderate financial means. Their establishments are extensions of domestic and working spaces, where customers spend a lot of time yet little money on a daily basis. They mostly appeal to digital natives, who use social media to communicate and disseminate information and get updated on recent trends. Nevertheless, these digital natives are less interested in high-end dining. Rather, they use these spaces as settings in which face-to-face interactions take place, tangible communities develop, and feelings of camaraderie are formed, all with people like themselves, who reaffirm their identity via consumption and use the setting as a platform for generating changes in the neighborhood.

Global Brooklyn Tel Aviv enables millennials to settle into jobs that were once considered working-class occupations, attaching cultural and symbolic capital and social appreciation to their work. Owners and employees often see these settings as enabling them to educate customers to acquire new tastes, identifying themselves with counterparts in other global cities. Yet, while they appreciate the knowledge and bodily work invested in producing foods and beverages, many regulars and visitors still prefer what they themselves consider the "wrong" taste of coffee, hummus, pizza, or wine. In other words, although exposed to "correct" knowledge, customers are less enthusiastic about the new tastes, but rather are appreciative of the presence of establishments that provide a global experience at an affordable price range and offer homemade food. In navigating between attempts to introduce a global concept and compromises to guarantee profitability, entrepreneurs instil different meanings into locality, together with good food and beverages.

Although it deviates from the ideal type of Global Brooklyn suggested in the Introduction to this volume, Global Brooklyn Tel Aviv reveals new forms of cosmopolitanism catering to the pride and passion for instilling different interpretations of the local and enabling the formation of active communities and consumption spaces in gentrified areas. In Tel Aviv, it is mostly the ethos of sociability, together with the call to make a moderate yet fulfilling living based on manual labor that sustains Global Brooklyn.

References

Bourdieu, P. (1984), *Distinction*. Cambridge, MA: Harvard University Press.

Bourdieu, P. (1986), "The forms of capital." In J. Richardson (ed.), *Handbook of Theory and Research for the Sociology of Education*, 241–58. New York: Greenwood Press.

Cronin, M., McCarthy, M., and Collins, A. (2014), "Covert distinction: How hipsters practice food-based resistance strategies in the production of identity," *Consumption Markets & Culture*, 17 (1): 2–28.

Greif, M. (2010), "Positions." In M. Greif, K. Ross, and D. Tortorici (eds.), *What Was the Hipster? A Sociological Investigation*, 4–13. New York: n+1 Foundation.

Halawa, M., and Parasecoli, F. (2019), "Eating and Drinking in Global Brooklyn," *Food Culture & Society*, 22 (4): 387–406.

Hubbard, P. (2016), "Hipsters on our high streets: Consuming the gentrification frontier," *Sociological Research Online*, 21 (3). Available online: http://www.socresonline.org.uk/21/3/1.html DOI: 10.5153/sro.3962 (accessed July 30, 2019).

LeBesco, K., and Naccarato, P. (2015), "Distinction, disdain, and gentrification: Hipsters, food people, and the ethnic other in Brooklyn, New York." In K. M. Fitzpatrick and D. Willis (eds.), *A Place-Based Perspective of Food in Society*, 121–40. New York: Palgrave Macmillan.

le Grand, E. (2018), "Representing the middle-class 'Hipster': Emerging modes of distinction, generational oppositions and gentrification," *European Journal of Cultural Studies*. Available online: https://doi.org/10.1177/1367549418772168 (accessed July 30, 2019).

Mali, I., and Varis, P. (2016), "The 21st-century hipster: On micro-populations in times of superdiversity," *European Journal of Cultural Studies*, 19 (6): 637–53.

Michael, J. (2015), "It's really not hip to be a hipster," *Journal of Consumer Culture*, 15 (2): 163–82.

Milkman, R. (2017), "Anew political generation," *American Sociological Review*, 82 (1): 1–31.

Mohar, Y. (2015), "The low-income college-educated," *Israeli Sociology*, 17 (1): 79–100.

Schiermer, B. (2014), "Late-modern hipsters: New tendencies in popular culture," *ActaSociologica*, 57 (2): 167–81.

Scott, M. (2017), "'Hipster Capitalism' in the age of austerity?," *Cultural Sociology*, 11 (1): 60–76.

Stahl, G. (2010), "Mile-End hipsters and the unmasking of Montreal's proletaroid intelligentsia: Or how a Bohemia becomes BOHO." Available online: http://www.adamartgallery.org.nz/wp-content/uploads/2010/04/adamartgallery_vuwsalecture_geoffstahl.pdf (accessed September 30, 2020).

Taylor, P., Doherty, C., Parker, K., and Krishnamurthy, V. (2014), *Millennials in Adulthood*. Washington, DC: Pew Research Center. Available online: http://www.pewsocialtrends.org/2014/03/07/millennials-inadulthood/ (accessed July 30, 2019).

Tuz, J. (2014–2015), "Hipsters and the city," *TransCanadiana: Polish Journal of Canadian Studies*, 7: 215–31.

Zukin, S. (2008), "Consuming authenticity," *Cultural Studies*, 22 (5): 724–48.

Dispatch

Accra: Who Is Eating in Global Brooklyn?

JT Akai

It is forty-five minutes past 5:00 p.m. in Accra. The clock is ticking as the white-collared workers of Accra Central saunter out their office buildings and flood the streets. Their eyes are lit as they take in the energy of the evening. They are well on their way, and thank God it's Friday. Friday nights are for the pleasure-seekers and what is better than starting the night with a hearty dinner and complimentary samosas, Global Brooklyn style?

Who is eating in Global Brooklyn? Just about every millennial and everyone else that can afford it. Seems like a large market niche? Over the past two decades, Ghana's infrastructural development has been soaring, and this has also made way for an expanding middle-income class. Although the middle class's expansion is not as rapid as it was envisioned by statisticians, it is still happening, generating new cultural dynamics.

Follow the string amber lights and you will find good bodega-style food. We are talking roasted pork, sizzling tawny to the rind, marinated with thyme and smoky leeks. Smoked salmon blistered with lemon and topped with snipped chives. Crab cakes cured with the right amount of coriander to complement the sauvignon blanc. Sounds overly refined and bourgeois? Can't be helped. Ghana's middle class has disposable income, and they spend on the finer things.[1]

It sounds almost absurd for a generation that reports of staggering unemployment and housing deficit rates (Baah-Boateng 2018) to still be desiring the finer things in life. While many may assume the radical hypothesis that Ghanaian millennials have misplaced their priorities, I, as a Ghanaian millennial, would beg to differ. It is my belief that Ghanaian millennials who have been exposed to the magic that is internet globalization through social media have become self-assured of how fleeting every moment is. The best way to unwind and purge is to enjoy good food, cathartic music, and the spectacle of a chill ambience.

In the next few pages, we will take a journey through four of Ghana's prominent Global Brooklyn-style restaurants. This will help us on our discovery to determine the type and caliber of people patronizing these places. We start with Lé Must, because if it must be done, then it must be done well.

Lé Must

The hubbub in this place is dim, just like the lights. The walls are covered with mahogany and lined with booths. The spaces in between are clustered with square tables and chairs. This is Lé Must, the restaurant in the square of the Accra Mall. Do not let the name dissuade you. Trust that this restaurant is not a French one, or at least its menu and vibe are not. The atmosphere in this place is slow and soothing, almost like in a jazz club. Just the kind of place where you know you can have a conversation over the table without much care for who is eavesdropping. The aesthetic appeals to your indulgence, and after a long and hard week, it is one of the best places to wine, dine, and unwind.

The service here is as professional as anywhere else internationally. The owners of Lé Must know their market niche well enough, as one can easily tell by their menu and their all-round service. It is almost elementary, considering where the restaurant is situated. The Accra Mall is without a doubt one of the busiest fixtures in the country. Situated to the east of the Tetteh-Quarshie Roundabout, the Accra Mall is placed in one of the most valuable pieces of real estate in Ghana. The Tetteh-Quarshie Roundabout intersects the vast and industrial expanse of Spintex and Tema (the east of Accra) and the sprawling residential space of west Accra, so one can certainly understand why the Mall would be such an eventful place. And as almost all malls go, they attract a youthful bunch.

Lé Must's décor is essentially industrial, which is noticeable of many Global Brooklyn bistros in Accra. As the country is still developing into an industrial phase, certain motifs, artistic or otherwise, which may seem near-historic and postindustrial elsewhere, are part of the present. However, with its concrete and exposed bricks, Lé Must also gives us a touch of wood to highlight traditional African aesthetics. This is where the clash happens: Western and African; sophisticated but traditional; different and yet familiar. The best people attracted by such motifs in Ghana are millennials. They seek variety and difference but still cherish their identity as a defined group of people. Lé Must's menu, which includes continental dishes and Mediterranean sauces, also doubles up with the best of the tropics in the form of fruity smoothies and cocktails. Lemon-infused mango juice accompanied with steaming-hot sautéed greens in beef sauce and vegetable rice is an absolute favorite.

Capitol Café and Restaurant

Capitol Café and Restaurant offers one of the best marinated steaks in town. The first thing that greets you here, after the gleaming white lights of course, are the scents of herbs and the sound of sizzling blocks of meat over a grill. The eating area is a wide tiled space and but for the scattered soft chintz chairs it would seem very sterile. Trust me, Capitol Café is anything but that.

The café is barely a ten-minute drive away from the American Embassy inside Cantonments. So, you already get a measure of the type of place it is and the people you will find in there: old money and nouveau-rich folk. Capitol Café is heavily modernist in its architecture, and this is certainly because of the people that visit there. However, the café also qualifies this style with wood paneling and armchairs.

While it remains posh and exquisite, the menu is miscellaneous in that it melds not just the Accra and Global Brooklyn lifestyles but also that of the owners, who are Lebanese. The café brings you Asian berry smoothies and chai teas alongside tropical fruit bowls. The kitchen is exposed to patrons, which is novel in Ghana. You get to savor the remarkable journey and fanfare that accompanies the resident chef's delivery of your meal. The heady smell of coffee and pistachio beans being blended, the smell of fruit pulp being churned into liquid delights, and the whiff of cumin and some mystery spice that speaks of sizzling hotness, are among the many sensory stimulations prominent in Capitol Café. From Italian-style pasta with cheddar cheese to deluxe burgers with a side of salty-sweet fries, there is something for everyone. This is not to say you will find just about everyone here.

This restaurant entices a seasoned group of patrons: the kind with income in six figures and over, the stock of people with an epicurean taste for fine dining. It is easy to find patrons of varied nationalities here; the executive American craving decadence, the affluent Lebanese curbing his family's want for a halal feast, and the Ghanaian highbrow seeking quality with every bite off the fork.

With a veritable aesthetic hankered on capaciousness, exclusive comfort, and an open-kitchen bustling with gourmet chefs, Capitol Café and Restaurant is a place not to be missed in Ghana's capital.

Zen Garden

Accra's road traffic is utter stress. In the early hours of the morning and the twilight of the day, the roads are choked. It is rush hour and everyone is all about their business. However, at happy hour you know where you will find white-collar workers, and that is Zen Garden.

Zen Garden inside Labone-Accra is the local man's place of zen. Which is quite funny considering meditation almost never includes acoustic music, kaleidoscope string lights, and hue-swirling umbrella-tipped cocktails.

This bar is one of those secular places you go to soothe your soul. Its aesthetic is literally lushness, with the greenest verdant plants in the capital and crisp air as the mixologist razzle-dazzles. The only thing louder than the jovial banters is the soul sister at the bar's dais performing spoken word and lyric rhymes. The whole bar is like a picnic inside the Brooklyn Botanical Garden and Arboretum. The lights are not too bright here; they shine and string from tree branch to tree branch like glowing gossamer. This is the place to let your hair down and immerse in the music's serenity.

The most impassioned theme, which makes Zen Garden so Brooklyn, is simply the soul of the place. It is a bar and grill, but for all intents, it is a shrine for music and the arts. With varied concerts and visiting top Ghanaian and Nigerian acts passing through, Zen Garden never leaves you underwhelmed. It is no surprise that the people you find attending Zen Garden's cantatas and acoustic sessions are people with love for alternative music and Afrobeats. The crowd here is mostly young, with a heavy concentration of American and European tourists. They seek a thorough helping of Ghanaian culture in a space, which they feel a sense of familiarity with; Zen Garden's lush soul provides that cultural environment. It is never a torpid night here, and the people that visit know this all too well. It is a hipster's paradise. The young crowds are eager to partake and the relatively older patrons, who are themselves adventurers of sorts, are ready to join in the fun.

Zen Gardens organizes some of the most conceptual concerts, and they do it very well. With acclaimed music acts such as Trigmatic and the high-life band Kwan Pa, Zen Garden is a holistic place to immerse yourself in true Ghanaian music. The place is a whole vibe on its own. They do not call themselves Serenity in the City just because of their plants.

Burger and Relish

Our final destination to eat at is Burger and Relish inside the backend of Osu's Oxford Street, Accra. Burger and Relish is officially Ghana's number one gourmet burger place, and their aesthetics reflect it. Just like almost all restaurants that have a Global Brooklyn vibe, Burger and Relish utilize small spaces in ways that create the illusion of them being larger than life. Their distinct theme is casual and more reminiscent of hipster culture. They have got streaming amber lights and wooden tables with an assortment of seating, which includes the typical American diner vinyl-seating booths, colorful metal chairs, and tall stools. The vibe here is easy, with Afrobeats pumping in the background. Sometimes a slice of hip-hop gingers up the diners.

The burgers here are pricey and that is only well, considering they are all deluxe-sized. Their buns are thick and grainy with sesame seeds, and their filling is dribbling with hot cheese and cayenne-spiced patties. Their side fries are, and I say this with reverence, the saltiest and spiciest I have ever had. They are dabbled with thyme, cayenne, and chili, and they taste like heaven with every chomp. The cocktails are the most exotic and fruitiest ever, with squashed passion fruits, pomegranates, and peaches bobbing in the tonic.

The fantastic thing about Burger and Relish is not just the food but the feeling of amity that seems to pervade this place. The owners organize a very mean Game Night every Thursday, and it summons some of the grittiest and most competitive scholars of pop culture. It is amazing. Considering the games, which are subject to change depending on the umpire's rules, the people that come here are a relative mix of middle-class millennials and middle-class foreigners. The latter you can recognize because they stick out like sore thumbs with their accents and their raucous laughter whenever they win which, surprisingly, is quite a lot, considering the games sometimes involve Pan-African history.

The people eating in Global Brooklyn-inspired restaurants in Ghana are miscellaneous in their age bracket, but they are typically high or middle-income earners. It is no surprise, considering they have the most disposable income and are prone to exercise both local and globalized gourmet sensibilities. With the expansion of social media and the resultant rise in taste-conscious consumers in Accra, the quirky decadence of these uniquely themed restaurants seems to be rubbing off on older restaurants in the country. They are all beginning to rebrand their bars and restaurants.

Global Brooklyn, with its motifs of dare-to-be-different and yet contemporary, is definitely something that Ghanaians have very well latched on to, guessing from the new themed restaurants which seem to be popping up all over the capital, food, and the entertainment that comes with heat, but one that our Ghanaian appetites are prepared to handle.

Note

1　Middle- and upper-class Ghanaians are increasingly using their purchasing power for luxuries as opposed to a few decades ago, when moneys were mostly spent on necessities.

Reference

Baah-Boateng, W. (2018), "Jobless growth is Ghana's biggest youth challenge," African Center for Economic Transformation, February 27. Available online: https://acetfor africa.org/highlights/jobless-growth-is-ghanas-biggest-youth-challenge/ (accessed February 27, 2020).

7

Mumbai

Importing and Glamorizing Social Values

Priyansha Jain

It was in late 2016 when I moved back to Mumbai after three years in London. London had been incredible for various reasons, but one of the most significant ones was its ubiquitous coffee shops. Living in the east of the city, I was always a short walk from specialist coffee shops like Ozone Coffee Roasters, Prufrock Coffee, Attendant, All Press, Barbour and Parlour, or Climpson and Sons, and frequented them very often. These coffee shops provided a relaxed atmosphere with chirpy staff, fast wifi, and accessible plug points, allowing me to work and socialize seamlessly while sipping on well roasted and brewed coffee. I was certain that on my return to Mumbai, I would not be able to procure a cup of flat white in similar environments.

Back in India, I was not looking for any mega coffee chain. Those kinds of places did not give me the *feeling* I was looking for, even if I could not fully understand why. And today, three years later, that *feeling* is an essential starting point to describe cafés in Mumbai that reflect the Global Brooklyn style, as discussed in the introduction to this book.

Blue Tokai, South Mumbai

In the early days of my return to Mumbai, I was due to visit G5A, a new contemporary cultural center in the southern part of the city. It was located in Laxmi Mills Industrial Estate, a compound built on the remains of old decaying textile mills. Interestingly, new history is being written over these mills, with independent design shops and art and cultural centers popping up in the vicinity. On my visit to G5A, I was surprised by the scent that filled the compound: it smelled like freshly roasted coffee!

The aroma was diffusing from an establishment called Blue Tokai. On the exterior, the floor-to-ceiling glass front gave away the plan of the interior: the space was divided into two parts by a slim wall, where an old aluminum *thali* (plate) was hung, used as a hand-painted signage along with some suspended moss ball plants. The area on the right had numerous jute sacks, stacked over one another, and was dominated by a gigantic coffee-roasting machine. On the other side, I saw wooden tables of different sizes and a counter with coffee machines and various coffee equipment. Stepping inside, each section could be observed from the other side through large circular glass windows that had been cut out in the wall. There was a sense of transparency in everything.

The room was filled with natural light and exuded a feeling of ease. The wooden furniture and the low counter were custom-made. Unlike in other Global Brooklyn establishments around the world, however, the furniture was neither old nor upcycled. Industrial-looking lamps hung from an old wooden ceiling that had been retained as a characteristic element from bygone times. The brand's primary and secondary colors—varying from aqua to turquoise— accentuated the white walls. The tones of blue created a sense of calm, along with the terrarium jars placed on the tables and the *kokedama*, the Japanese moss balls hanging throughout the establishment, creating a cocooned space amid the chaotic city.

Over the counter, I spotted familiar artisanal brewing equipment: La Marzocco, AeroPress, Chemex, and Hario Drip Kettle, all meant to extract the most optimal flavor from coffee that was being roasted in the other room, applying techniques meant to provide a flawless brew. The coffee was sourced from ten estates in the state of Karnataka, South India, and each variety was roasted at a specific temperature that brought out their most complex taste profile. The menu was printed on matte textured paper and listed coffee choices as distinct as espresso, flat white, iced pour over, and even nitro coffee (cold brew coffee charged with nitrogen)! There were options such as cow, almond, soy, and lactose-free milk, uncommon in India. The menu was dominated by coffee, although there were some bakery essentials like croissants, cookies, and chocolates, all sourced from artisanal brands like Mag Street Kitchen, based in Mumbai, and Mason and Co., operating from Pondicherry.

There was change in the air! The staff, warmly welcoming and expressing greetings to customers on their arrival, were wearing custom-designed denim aprons with the brand's logo embroidered on them, and the back of their T-shirt read "We love Indian Coffee Farmers." This is unseen and unheard of in India, where the century-old caste system still impacts people's behavior and hampers interactions, especially among different classes. I was amazed to notice the equal ratio of men and women working at the coffee shop. The service industry in the food and beverage sector in the country has been dominated by men for a long time; a traditional reasoning for this has been the erratic working hours that are not supposed to suit the lifestyle and safety of women. Taking note of fellow customers, I realized there were many tables with a sole occupant, a rarity in India's food and beverage culture where eating out is typically a group activity.

The staff in India is often overlooked, as there is an abundance of available labor. However, since a café such as Blue Tokai requires skilled labor to operate coffee machinery—*makers* in the process of roasting and brewing single-origin coffee—staff members are absolute key here. They are trained to discern the different temperaments of coffee and taught to understand the complexities of the craft. In contrast to other countries, staffers in Mumbai are often recruited as waiters and then trained as baristas. Since third-wave coffee is still in its nascent stages in Mumbai, there are no skilled baristas available on the market. The passionate owners and founders of cafés, who often tend to be of foreign origin or have lived abroad for many years and are well versed in café culture, are often seen personally training the staff. Many staff members at these establishments then gain enough experience to be able to train new recruits.

The transmission of knowledge is not only limited to the staff but also extends to its customers. Looking at Blue Tokai's social media account, I am impressed by the frequency of the events directed to their clients. Free sessions like Intro to Roasting, Intro to Coffee Cupping, Bean—Brew—Bar, Fundamentals of Coffee, or Filter Brewing take place weekly. Blue Tokai not only tries to establish a relationship with its customers but also thrives on associating itself with local brands such as Korra, a jeans brand that keeps away from mass manufacturing and whose every pair of jeans is stitched by one tailor from start to finish, and Bombay Underground, which makes zines as a form of art, expression, and social exchange. The overlapping ethos of all the brands Blue Tokai collaborates with emphasizes local production over imports, small batches over mass production, and a strong focus on community engagement and sustainability. Within its coffee shop, Blue Tokai holds pop-ups that are a day or two days long, organized together with underground zine clubs, artisanal perfumes, fair-trade fashion studios, and offbeat travel companies. In extending its welcome to accommodate these various other practices, the café aligns itself as an alternate space and creates a community of like-minded people around it.

A Short History of Coffee (and Tea) in India

The success of a coffee shop like Blue Tokai is especially interesting as India is usually perceived—erroneously—as a place exclusively focused on tea. It cannot be denied that tea is ubiquitous in India, an everyday element of households and social gatherings regardless of caste, class, or religion. As a consequence, the Global North often sees India through the "chai-latte" lens. In 2012, tea came close to being declared the National Drink of the country (Press Information Bureau 2012). Since the twelfth century, tea was found in the mountainous areas of today's Assam as a wild plant that was used by the local tribes for its medicinal properties. In 1820 the British finally found the plant and started cultivating it in the Northeast of India (Nelson 2012). Forward thinking and vigorous advertising campaigns through the twentieth century for both colonial and national trading purposes popularized chai and made it the preferred beverage (Lutgendorf 2012).

Contrary to common misconceptions, in India coffee was cultivated before tea. A famous legend narrates that in the late seventeenth-century Sufi Saint Baba Budan strapped seven coffee seeds on his chest and smuggled it out of Mecca, Yemen, as the export of fertile coffee seeds was at the time prohibited by the Arabs. On his return, he planted it in Chikkamagalur, Karnataka (Wild 1994). In the mid-nineteenth century, when the British colonial presence took strong foothold, coffee plantations expanded for export.

Coffee in India is a shade crop and grows under rich vegetation that is area-specific, including silver oak trees, black pepper vines, and orange trees, and amid various animals and birds (Packel 2010). Two species of coffee have been produced in India: Arabica and Robusta, mainly grown on the hills of Karnataka, Kerala, and Tamil Nadu, three States in South India. Farmers in India prefer the Robusta trees to Arabica even though they yield lower prices as they are stronger and resistant to coffee rust, a fungus that Arabica trees attract (Allison 2013). Following India's independence, quantity and not quality became the focus between the 1940s and the 1990s, when the Union Government pooled the coffee from different provenances, marketed it through the Indian Coffee Board, and paid the farmers by the pound of product (Allison 2013).

In the coffee-growing parts of South India, a drink called Indian Filter Coffee became popular. Served in a stainless steel tumbler and a saucer with high side walls and a lip, it was made with dark-roasted coffee beans and chicory, yielding the taste of a milky sweet coffee. At the time, the understanding of coffee in most other parts of India was limited to the instant kind, served with milk. In 1936, the Indian Coffee Board opened the first India Coffee House in Churchgate, Mumbai. It was a café serving coffee and South Indian food. By the 1950s, it had seventy-two outlets across India, familiarizing the tea drinkers of the North with coffee (MJ 2011). These cafés were painted simply and usually devoid of any decorations; the waiters wore turbans and cummerbunds (broad waist sashes) that added a royal touch; they were a common ground for lawyers, politicians, artists, academicians, and students alike (Krishan 2016).

In the 1960s, when India had familiarized itself with both coffee and tea, cheaper production methods of tea had considerably lowered its price (and quality), leading to the emergence of *chai-wallah* (tea shops) on the streets (Nelson 2012) while popularizing tea and making it available to everyone regardless of class. With the liberalization of the Indian economy in the 1990s, producers were incentivized to brand coffee individually and cafés that aspired to reflect emerging forms of novelty started opening.

Café Coffee Day was one of the first modern cafés where guests were introduced to coffee styles other than South Indian Filter Coffee or instant coffee, namely espresso, latte, cappuccino, and Americano. It was a safe space (with "right of admission" reserved) that provided air conditioning (Lutgendorf 2012), allowing one to loiter around for as long as one wanted by merely ordering one coffee. Even though the beans used were lower grade Robusta, for the middle class it was an aspirational space that provided them with a chance to experience something "Western." For college students, it was an opportunity to hang out or partake in romance, and for businessmen a place to make deals. For some it was even a space to fix marriages or finalize divorces (Biswas 2019). Additional toppings of cream and flavors like hazelnut and caramel, along with the convivial experience of drinking coffee in a public place, convinced guests to pay premium prices. Today, Café Coffee Day has around 2,000 branches in India, opened strategically close to colleges, offices, and residential areas. Barista and Costa Coffee are other noteworthy second-wave coffee shops.

A decade later, in late 2012, Starbucks opened its first Indian branch in Mumbai. The coffee shop is situated in a historic building occupying two floors with interiors reflecting Indian craftsmanship and patterns. Along with various coffee offerings and preferred flavors and toppings, one can indulge in the savory and sweet food options that reflect an Indian palate. Starbucks in India is a joint venture with Tata Global Beverages Limited, the world's second-largest manufacturer and distributor of tea and a major producer of coffee. Through this partnership, the American company gets access to sought after properties, along with a direct relationship with coffee farmers. As of 2020, Starbucks has a hundred stores across six major cities in India.

Today, with the arrival of third-wave coffee shops like Blue Tokai and Koinonia in Mumbai, one still observes a dependence on tea as a way to move forward with coffee. In 2016, Blue Tokai uploaded a video on their YouTube channel (Blue Tokai 2016), which explained how one could make coffee with freshly ground beans by using a tea strainer commonly found in any Indian household.

Koinonia, North Mumbai

Koinonia first opened in early 2017 in a narrow lane in the quaint area of Chuim, in the suburbs of Khar and Bandra, in the heart of Mumbai. It is a neighborhood frequented by expats and nonresident Indians; the descendants of the Portuguese Christians also live here. It is a trendy neighborhood that attracts young entrepreneurs interested in engaging with the entertainment, hospitality, and creative industries. The historic streets with the charming Portuguese style bungalows, unlike the surrounding urban landscapes, make you feel like an explorer.

Chuim Village can be a maze and often even Google Maps does not help one locate Koinonia; the solution is to surrender to the smell of freshly roasted coffee and follow it. The black façade of the café is easy to spot, as is the statue of Jesus Christ on the front of the coffee shop. Koinonia means "Christian fellowship or communion" with God or more commonly with fellow Christians, and the Australian founder, who is of Indian descent, believes in creating a community among like-minded individuals who are interested in coffee.

Inside, Koinonia is a small, 250-square-foot café; one half is painted black and features a Probat roasting machine and the other half, with protruding wooden stripes, is painted white. The furthermost wall displays a hand-painted logo featuring bold typography with exaggerated letters reading "Koinonia Coffee Roasters," along with a cappuccino cup with a little sailing ship afloat. An inspiration for the logo could possibly be the neighborhood village and the Portuguese explorers who arrived in the city in the early sixteenth century.

On most days, one is greeted with rich cocoa- and molasses-filled air. As a regular at Koinonia, one can immediately point out the origins of this fragrance: beans from an altitude of 5,000 feet from the Marvahulla Estate, on the Nilgiri Mountains in the state of Tamil Nadu, South India. The design of the counter presents two sides at different levels: the side facing the barista, where coffee is prepared, is lower and stocks a La Marzocco espresso and grinding machine. The side of the counter facing customers is elevated, creating a wooden bar with sleek bar stools and suspended lights that echo an industrial look. Other familiar equipment like the AeroPress, Clever Coffee Dripper, Siphon, Chemex, and HarioBuono Kettle are dotted around the café either being used or available for sale. Along with the usual third-wave coffee offerings, an *affogato* is available as a local addition to the menu to beat Mumbai's heat. Unusual and interesting ice cream flavors are served with the *affogato*, like "Dark Chocolate Italian Truffle" from a local ice cream shop called Bono Boutique. Cashing in on the ongoing trend prevalent in the city, the café is also offering "Keto Coffee by Coach Urmi," that consists of organic coconut oil and organic filter coffee sprinkled with organic cinnamon. Coach Urmi runs Kinetic Living, a reputable fitness and wellness studio in the neighborhood. In one corner of the café, one finds high-quality light-, medium-, and dark-roast single-origin coffee, packaged and available for sale. The coffee comes from various South India coffee estates like Marvahulla Estate, Doraikanal Estate, and Poabs Estate. The slick black packaging for the coffee and the takeaway cup is an extension of the stylish café space. The cooling tray of the roasting machine has been modified to allow for the addition of a tabletop, accommodating two more guests of the total twelve that the café can seat at any given time.

The cozy space makes it nearly impossible for guests to not overhear each other's conversations; returning customers often end up turning into acquaintances and often friends. On a normal day, a set of typical guests who are regular at Koinonia may include a part-time Polish painter and full-time line producer for Bollywood, a Swedish-Spanish man teaching French at the French International School of Mumbai, an Indian-British couple from the theatre world who have recently moved to Mumbai, a local interdisciplinary designer who studied in London, a Bollywood producer, an independent filmmaker, a chef who spent his childhood in Bandra but now lives abroad, and a local from Bandra who has never left the city. Some guests come to enjoy a cup of distinctive coffee, some to catch up with friends, and some to conduct work meetings. Some come along with their laptop, boosting their productivity with the background noise, whereas for some, it is just a part of their daily routine.

Koinonia also employs a full-time "community manager" whose job is to welcome guests and introduce them to the culture of coffee. That entails explaining to the locals what a "flat white" is and how it is different from a latte, while also encouraging them to taste a pour over. To regular customers and ones who are well versed in diverse types of coffee, the community managers would describe the farms in relation to the bean and coffee of their choice. One of their other significant roles is to remember the preferences of the regular guests, ranging from preparing piping hot coffee to serving latte in a glass instead of a cup or playing a specific playlist.

The Australian owner's ubiquitous presence in the café and his thorough involvement with staff and customers is a rare sighting in India's food and beverage scene. The founder often hires young adults (eighteen to twenty-four) from orphanages through an agency and trains them to understand coffee and become baristas. Since the third-wave coffee scene in India is still nascent, it is rare to find a well-trained barista on the job market. However, applicants with interest in coffee and recent graduates from culinary courses or hospitality schools might also apply. Once they have trained at Koinonia for a specific period and once they demonstrate a certain skill level, they are considered for the title of senior barista, after which they can be put in charge of training new employees. The founder also tries to take the baristas to the estates in South India where Koinonia procures its coffee, helping them embrace a holistic way of learning that reflects the company's ethos.

The founder brings with him a work culture he learned abroad, trying to inculcate more respect for each member of the staff than what is common in India. However, the coffees themselves would be prohibitively expensive for the baristas. This is a manifestation of enduring class differences, a consequence of the caste-based system that is still practiced in India. In a local *udupi*, which in Mumbai refers to simple eateries, considered less "trendy" but still serving delicious food and often brimming with people, one would experience an extremely vertical relationship between patrons and staff as it is taken for granted that the employees are present to "serve" them. Customers visiting Koinonia are usually more aware of social hierarchies and hence more likely to establish a horizontal relationship with employees. This is visible when patrons often return the staff's pleasantries and ask them about their well-being.

While a café like Koinonia would seem to present many of the characteristics of a Global Brooklyn café as highlighted in the introduction of this volume, its business model is already starting to take a different turn, just a few years after its opening.

Manifestations of a Global Brooklyn Café: Scaling and Glamorizing

Paying rent in trendy areas and importing specialized coffee equipment makes it quite challenging to sustain the café business. In these scenarios, Blue Tokai and Koinonia have found various ways of adapting the Global Brooklyn coffee models to face the challenges of the market in Mumbai and to generate profit.

In 2016, through seed financing, Blue Tokai was able to open multiple cafés across India in New Delhi, Mumbai, Bangalore, Gurgaon, Noida, Goa, and Jaipur. Within Mumbai, they have seven of their current total of twenty-two cafés in India. Such proliferation of new branches is in stark contrast with the way Global Brooklyn operates in the Global North. However, by multiplying the cafés, the Blue Tokai entrepreneurs are able to reach more people and educate them about India's single-origin Arabica coffee—a shift from the instant coffee Robusta mix. From one neighborhood to another, it is essential for them to adapt their brand so it can relate to site-specific customers. For instance, one notices that locations for Blue Tokai cafés in trendy and artistic neighborhoods like Mahalaxmi and Bandra tend to be in narrow and hidden lanes; the signage and logo are small and more discreet as well. Instead, in other more residential or commercial areas the cafés tend to be located on the main roads with signage and a logo that are much bigger and much louder. The Blue Tokai cafés in Mumbai open as early as 7:00 a.m. and close as late as 12:00 a.m. These opening hours are convenient for people working in formal jobs and informal jobs (taking care of children and managing homes as full-time housewives, for instance).

Koinonia on the other hand has also taken measures to reach the masses by selling on Amazon, just like Blue Tokai. But to raise its profile and make the brand easier to remember on the platform the owners took the decision to change the name of the café from Koinonia to KCROASTERS (KC standing for Koinonia Coffee), although the complete name on their Amazon Profile feels a bit repetitive: Koinonia KCROASTERS. While undertaking these changes, they also rebranded, getting rid of the unique cappuccino cup logo with the floating sailing ship. Their new "KC" monogram logo is encased in a box that gives out a very corporate feeling. Abstract color illustrations have been used on the packaging, with a different color distinguishing each estate. However, the new branding lacks a relationship to its original physical space or neighborhood, creating a very generic mood.

In 2019, KCROASTERS opened at FoodHall's in Khar, within ten minutes distance from its first Chuim Village presence. FoodHall is a gifting supermarket with a very high-end feeling to it, conceived as a one-of-a-kind store for special occasions, where one can pluck herbs from a vertical garden or get the most exquisite international groceries. FoodHall has partnered with esteemed tea, chocolate, and coffee brands so they can have an outpost at its 25,000-square-foot shop spanning over four floors. One could say that KCROASTERS presence at FoodHall is antithetical to the look and feel at its Chuim location. It has fitted into the molds cast by the luxury supermarket, with brass lettering used for the signage "The Coffee Lab" and "KCROASTERS," as well as glass panels demarcating the space. Every square inch of the floor area is lit up, and four siphon machines and another Probat roasting machine are the center of attention. Coffee from international estates in Africa and Guatemala are being featured here, sold at seven times the price of the Indian coffee. The owners capitalize on occasions like Valentine's Day by preparing mixes like the "Valentine Blend" with predictable dark chocolate and strawberry notes. Packaged coffee and equipment available for sale are at the forefront, unlike at their Chuim Village location where they are tucked away in a corner. Three elevated wooden bar counters, along with minimalistic bar chairs surrounding the main counter, remind one of a takeaway store. The space is designed to drive consumption and to create a memorable experience, with no particular intention to establish a community and so defying the original meaning of the word "koinonia." However, its presence at FoodHall still helps drive the third-wave coffee culture in Mumbai and allows for the richest of the rich to be able to experience single-origin coffee.

Conclusion

Global Brooklyn-style cafés are still a very limited phenomenon in Mumbai, both in terms of actual numbers and of the clientele that has the economic and cultural capital to afford and enjoy them. Their very manifestation in India is a luxury, as they are available and accessible to a minuscule part of the population. Global Brooklyn-style cafés become part of the local dynamics of social distinction, at times overlapping, at times resisting the traditional stratification, as they are usually located in upscale but open-minded neighborhoods. They refer to sensory environments and flavor profiles that are imported, but they adapt them to the local context. We have observed this in the custom-made furniture as well as the coffee infused with cinnamon.

As the Global Brooklyn cafés are not in partnerships with the big conglomerates of India, in order to survive, they need to expand, often embracing very traditional business models. This takes the form of numerous branches, retailing on Amazon, and opening within luxury supermarkets.

Global Brooklyn cafés showcase local Arabica coffee with a 100 percent Arabica sign on their packaging, playing an important role in supporting local Indian farms. The statistics available from the Coffee Board of the Government of India show a steady increase in the domestic coffee consumption between 2000 and 2011 (Indiacoffee 2014). This seems promising for the Global Brooklyn-style cafés and will hopefully encourage more to open in the next decade, eventually allowing for more independent and less corporate business models each passing year.

References

Allison, M. (2013), "As India gains strength, so does its coffee," *The Seattle Times*, January 27, 2013. Available online: https://www.seattletimes.com/seattle-news/special-reports/as-india-gains-strength-so-does-its-coffee/ (accessed December 6, 2019).

Biswas, A. (2019), "Emergence & economy of coffee shops in India," *Business-standard.com*. Available online: https://www.business-standard.com/article/opinion/emergence-economy-of-coffee-shops-in-india-119080801742_1.html (accessed November 1, 2019).

Blue Tokai. (2016), *Channi Coffee*. [video] Available online: https://www.youtube.com/watch?v=m2O5MqXD4n0 (accessed November 3, 2019).

Indiacoffee.org. (2014), *Consumption in India*. Available online: https://www.indiacoffee.org/coffee-statistics.html?page=CoffeeData#cof (accessed November 20, 2019).

Krishan, S. (2016), "When Indian coffee house was the country's living room," *Condé Nast Traveller India*. Available online: https://www.cntraveller.in/story/when-indian-coffee-house-was-the-countrys-living-room/#s-cust0 (accessed December 1, 2019).

Lutgendorf, P. (2012), "Making tea in India: Chai, capitalism, culture," *Thesis Eleven* 113 (1): 11–31.

MJ, R. (2011), "Good old Coffee House," *Tribuneindia.com*. Available online: https://www.tribuneindia.com/2011/20111113/spectrum/main4.htm (accessed November 4, 2019).

Nelson, D. (2012), "India claims to reclaim tea as their national drink," *The Telegraph*. Available online: https://www.telegraph.co.uk/news/worldnews/asia/india/9260322/India-claims-to-reclaim-tea-as-their-national-drink.html (accessed November 26, 2019).

Packel, D. (2010), "Coffee plantations in India blend history and hospitality," *nytimes.com*. Available online: https://www.nytimes.com/2011/01/02/travel/02coorg-explorer.html (accessed November 4, 2019).

Press Information Bureau. (2012), *Representations Received Declaring Tea as National Drink*. Available online: https://pib.gov.in/newsite/PrintRelease.aspx?relid=83943 (accessed November 25, 2019).

Wild, A. (1994), *The East India Company Book of Coffee*. London: HarperCollins.

Part III

Back to Brooklyn

8

Brooklyn

Hipster Aesthetics, Foodways, and the Cultural Imaginary

Kathleen LeBesco and Peter Naccarato

As the chapters in this volume make clear, the New York City borough of Brooklyn has come to be associated with a particular aesthetic that travels via global communication flows, embodying specific forms of knowledge, reflecting shared priorities and values, and promoting a particular ethos of production and consumption, especially where food and drink are concerned. However, our research reveals that the elements that comprise the transnational phenomenon of Global Brooklyn are closely aligned with specific neighborhoods, including those identified as hipster enclaves. In these spaces, highly stylized young people embrace social practices that are both celebrated for challenging conventional ways of living and vilified as the ridiculous indulgences of those with social and economic privilege. Regardless of one's attitude toward it, this hipster aesthetic—from fashion to interior design; from recreation to decoration; from restaurants to cocktail bars—has clearly emerged as a cultural export that marks a particular kind of urban space and citizen, and seeks to provide a recipe for living. In this chapter, we assess Brooklyn's place in the cultural imaginary by considering the similarities and differences between Brooklyn, the New York City borough, and Global Brooklyn. In doing so, we examine which people and practices are foregrounded and which are erased in these newly gentrified spaces and consider if such erasure is part of what gives Brooklyn its global traction.

While the aesthetics, practices, and values of Global Brooklyn extend beyond hipster culture, they are deeply influenced by it. So, the first question to ask is the following: Who are hipsters and how are they connected to Global Brooklyn? Put simply, a hipster is someone who self-presents as a sophisticated resistor to mainstream culture but who is perceived by others either as a caricature to be vilified or as an independent thinker whose do-it-yourself lifestyle is the hallmark of a successful countercultural identity. Hipsters tend to cluster in "hipster neighborhoods," which MoveHub, a company that provides local and international moving-related research, data, advice, and services, has defined as those with a requisite number of coffee shops, microbreweries, thrift stores, vintage boutiques, vegan restaurants, record stores, and tattoo parlors per 100,000 residents as well as rent inflation over the last year (Angst 2018; O'Brien 2018). In addition to featuring common types of businesses, these hipster neighborhoods also share common aesthetics. According to Juliet Carpenter and Loretta Lees, visual clues that indicate hipster gentrification include wrought iron, wood beams, exposed brick walls, cobblestone walkways, and interior courtyards with garden (1995: 299). With regard to interior style, be on the lookout for "Edison bulbs with orange-glowing filaments, recovered timber for ceilings, polished concrete, metal cabinets, randomized furniture, idiosyncratic bricolage: all nearing parody status now" (Bayley 2017: np). Stephen Bayley muses, "I found a marvelous business called Brooklyn Fabrication, which could furnish your entire loft in urban flotsam, retrieved lumber and bespoke metal furniture during a single weekend. Architectural salvage is Brooklyn's second language" (2017: np). As hipsters occupy these spaces, they have material and spatial effects that profoundly impact the cities they inhabit (Cowen 2006: 22 in Maly and Varis 2016: 648). And as neighborhoods change block by block, with unique bars, cafés, restaurants, and art spaces moving in and quickly replicating, this hipster style brings with it a focus on "good living" that has a transformative effect on cities (Maly and Varis 2016: 648). Certain features of this lifestyle—a do-it-yourself ethos, an artisanal embrace of the inconvenient— serve as a critique of other forms of "good living," including the materialism that characterized the yuppie culture of the 1980s. However, both of these lifestyles are negatively framed by the mainstream.

These markers may signify "hipster" in Brooklyn, but when they are found in other parts of the world (Global Brooklyn), they may not necessarily be read in relation to this culturally specific form of the Brooklyn hipster; what is labeled as "hipster" differs contextually depending on who uses the concept and in which part of the world (Maly and Varis 2016: 638). As Maly and Varis explain, "Hipsters in Ghent are not exactly the same as hipsters in Toronto or New York. There are several centers of normative orientation within hipster culture, producing different indexicals and identity discourses" (2016: 644). Although this Global Brooklyn aesthetic is linked to a particular type of hipster in Brooklyn itself, this may not be the case elsewhere in the world, depending on the shifting meaning of "hipster" globally. Nonetheless, as the chapters in this volume make clear, the aesthetics of these Brooklyn neighborhoods have traveled globally. As they have done so, they have circulated specific ideologies with regard to consumption and production that both promote and are promoted by economic shifts, including gentrification, that significantly impact the neighborhoods they occupy.

This process is better understood within the context of a postindustrial economy impacted by globalization, where individuals find themselves searching for "authentic" experiences and identities. One option for those with the requisite economic and cultural capital is to inhabit urban spaces where they can embrace a set of consumption and production practices through which they attempt to reassert their independence and recreate their identities. It is certainly the case that the hipster enclaves of Brooklyn value "authenticity," though definitions of this term are varied and contested. Nonetheless, authenticity is always understood positively, particularly during periods of great income inequality, when symbolic distinctions are used to articulate class positions (Finn 2017: 123). With regard to consumption, Brooklyn's embrace of authentic and hodgepodge, rather than sleek and artificial, offers opportunities for the kind of romantic consumption Colin Campbell describes—that is, consumption that is "imaginative, remote from experience, visionary, and preferring grandeur or passion or irregular beauty" (1987: 1). Such consumption produces a form of cultural capital that shapes not only these Brooklyn neighborhoods but cities around the world: "Those with social power have a monopoly over ways of seeing and classifying objects according to their criteria of good taste. The ability to create new systems of discernment is class power" (Bridge 2006: 1966). While "monopoly" might overstate the case, leaving no space for resistant criteria, the systems of discernment that result in the Global Brooklyn aesthetic do suggest some extent of class power at work.

As this is happening, we are witnessing a particular realignment of urban economies toward "cultural consumption (of art, food, fashion, music, tourism) and the industries that cater to it, [which] fuels the city's symbolic economy, its visible ability to produce both symbols and space" (Zukin 1995, cited in Ray 2016: 20). This process is supported by a landscape in which people's jobs—increasingly monotonous and bureaucratic—provide fewer opportunities for creative, independent expression, causing them to seek it out in their craft consumption (Campbell 2005: 38). Enter the hipster, who "emerges as a credentialed cultural producer who—following shifts in employment structures, ongoing state austerity policies, and digital revolutions—turns to creating micro-enterprises trading on everyday aesthetics, taste and style" (Scott 2017: 65–6).

Although hipsters may want to position themselves in opposition to a culture addicted to conspicuous consumption, Ico Maly and Piia Varis remind us that hipsters are not anti-consumption; rather, they promote a specific niche consumption market that fits within a broader economic infrastructure (2016: 646):

> Hipster culture is connected to a certain ethos of consumption, and from this perspective is perfectly aligned with the neoliberal structure of the world economy where, in our post-Fordist era, mass production for all has been replaced by niched mass production catering for the (identity) needs of specific, smaller groups. This niched production does not only sell products, but a mythology in the sense of Barthes (1957), and this can also be in the form of a countercultural identity. (2016: 648)

We contend that whether or not specific manifestations of Global Brooklyn are identified with hipster culture, they share in common this particular ethos of consumption.

As we come to understand these consumption practices in relation to the contemporary, postindustrial economy that produces Global Brooklyn, we must also recognize their historical roots. For example, the displacement that is the inevitable consequence of gentrification is not new; in fact, Marc Linder and Lawrence Zacharias (1999: 4) note that between 1880 and 1910 great swaths of agricultural farmland and countless farmers in Brooklyn were displaced to make way for industrial development and urban expansion. Perhaps the renewed emphasis on craft consumption in Brooklyn is a nod to this agricultural history. Linder and Zacharias (1999: 7) argue that "the result of urban sprawl has been a renewed rigidification of the historical division of labor between city and countryside, dichotomizing farmers and urban residents in both geography and attitude." It strikes us that the Global Brooklyn phenomenon in some ways attempts to bridge that divide. Linder and Zacharias (1999: 296) acknowledge as much when they note a trend toward urban agriculture "that confounds planners' dichotomous conceptualization of land-use boundaries," pointing toward the very urban farming initiatives, greenmarkets, and community gardens that characterize Global Brooklyn spaces.

Such an assertion pushes us to consider the extent to which the Global Brooklyn values that inform decisions vis-à-vis consumption also lend themselves to a similar ethos of production. Following Krishnendu Ray (2016: 20), we recognize that "the stark divide between production and consumption is no longer productive, especially in the articulation of taste in the American city."

As we consider the lived experiences of Brooklyn's aspiring entrepreneurs, we find a preference for (what is presumed to be more authentic) craft and artisanal production rather than commercial or industrial production. But as with the consumption practices discussed above, this contemporary celebration of craft and artisanal production has a particular history. Specifically, it is "the product of three centuries of urban transformation, a palimpsest of different 'waves of production' that reshaped the city and tied it to the changing international economy" (Warf 1990: 78). Entrepreneurial hipsters and the enthusiasts who embrace craft and artisanal production in their businesses are connected to this history in complicated ways.

Many of the neighborhoods in which craft breweries have become a telltale sign of hipster gentrification once hosted booming commercial brewing enterprises: Hipster neighborhoods Williamsburg and Bushwick have long been known for commercial breweries founded by German immigrants (Warf 1990: 80). In fact, as a result of German immigration in the mid-1800s, by the 1890s, there were forty-five breweries across Brooklyn's neighborhoods, more than in the cities of Milwaukee, Detroit, and the District of Columbia combined (Anderson and DeSena 1979: 130). But by the 1970s, there were no breweries left in Brooklyn until a new wave of craft breweries began to emerge in revitalized hipster neighborhoods.

Part of this revitalization depends on a shift in how these spaces are occupied. While the old breweries were exclusively sites of production that were inaccessible to the drinking public, these new craft breweries are communal spaces that blur the boundary between production and consumption. One effect of this shift is to invite consumers to engage directly with producers, likely contributing to the development of a discourse around artisanal beer that distinguishes it from what is perceived to be its inferior, industrially produced predecessor. For example, the craft beer consumer often critiques commercially produced beers as watery, thin, or tasteless, while celebrating artisanal beers for their individually selected malts, their unique ingredients and flavors, and their spectrum of colors and "tasting notes."

Thus, this re-emergence of breweries in Brooklyn is not merely a return to the past; rather, it is the result of significant social and economic changes that occurred in the interim. It also signals shifting values with regard to the production and consumption of beer (and other newly celebrated craft and artisanal products) and a revaluation of service as a means for creating and sustaining meaningful relationships between producers and consumers. As Ocejo (2017: 13) explains, "What separates the new cultural elites from their older peers and makes them part of a new form of luxury is how their workers intertwine interactive service with cultural knowledge and omnivorous tastes, and highlight a sense of craft in their work." This is certainly something that we encountered in our site visits. For example, Covenhoven is a bar on the edge of Crown Heights that is named after the farm that used to be on the site it now occupies. One of the wealthiest neighborhoods in Brooklyn at the turn of the last century, Crown Heights was dogged for years by a reputation for intolerance and violence after riots in 1991, but it is now experiencing significant demographic changes, with the black population shrinking to 70 percent from 79 percent from 2000 to 2010, and the white population almost doubling to 16 percent (Gregor 2015: np). Covenhoven is a cozy and inviting bar (for the newcomers, anyway, perhaps less so for the longtime residents who might not be in its target demographic) that describes itself as "your friendly neighborhood bar for beer geeks" that "champions the best and the brightest in local and regional brews."[1] Its staff includes beer manager Tom Beiner, described as "a teddy bear who wants nothing more than to share his latest beer scores." The website goes on to explain that "Tom's main mission is to source the latest and greatest beers and ciders available in Brooklyn."[2] Inside, the bar is lit almost entirely by candles on the tables, with very dim overhead lighting and the glow from the massive beer refrigerator in the back. One long wall of the bar features antique beer trays, including several for Schaefer, one of the last commercial breweries in Brooklyn to close before the recent resurgence of craft brewing. Under the beer trays are what appear to be wooden church pew benches and along the other long wall is a whimsical and abstract black-and- white mural, with the words "Learn to Swill" running nearly the length of the bar. Despite gestures to Brooklyn's brewing history, including the collection of antique trays, this is clearly not your father or grandfather's corner bar. This is craft beer with a hipster edge, signaled not only by the bar's aesthetics but also by its embrace of the forms, specialized knowledge, and skills that are the hallmark of artisanal production and service.

Such values and aesthetics are also prominent at 61 Local, a restaurant on Bergen Street, between the gentrified neighborhoods of Cobble Hill and Boerum Hill. With both its menu and its website announcing "We ♥ makers," its embrace of artisanal production is clear. The décor displays many familiar Global Brooklyn features, including exposed brick, Edison bulbs, and several long, communal wood tables. Wall sconces against the exposed brick are illuminated squares, covered with light, textured fabric and embroidered with any number of images (including a landscape of cupcakes, and a man on a bicycle). There is a wall laden with jutting iron racks that are draped with newspapers—very "old media." The cover of the menu underscores another one of the restaurant's defining values: local sourcing. It features a map of the mid-Atlantic and New England states dotted with the logos of a number of the restaurant's regional producers. This commitment to local sourcing is also highlighted on the website, which explains their philosophy: "We offer a locally-sourced menu that features premium products from passionate people. We take the time to visit with our producers to understand what makes their products unique and exceptional. By keeping it local, we're able to share this opportunity with our patrons, too."[3] The menu features some hipster classics, including deviled eggs, buffalo cauliflower, Kimchi'd kale salad, and avocado toast. At the bar, patrons will find "30+ tap lines pour[ing] a rotating selection of the best local beers, wines, and specialty craft drinks." The website invites you to "Come by and let our friendly and knowledgeable bartenders introduce you to your new favorite drink."[4] This embrace of local sourcing, specialized knowledge, and artisanal production ultimately connects to the overall focus at 61 Local, community: "61 Local is a true public house where the community comes to eat, drink, celebrate and collaborate. . . . It's about people; it's about connecting; and it's happening right now, over delicious food and drink, here at 61 Local."[5] Clearly, such values—and the curated experiences they produce—are meant to stand in stark opposition to the alienating effects of the modern industrial food system.

Although businesses like Covenhoven and 61 Local situate themselves in relation to a romanticized past, they are also implicated in processes of gentrification that are shared by the various sites that coalesce under the heading of Global Brooklyn and that have a common impact: "Jobs are lost and businesses closed because new residents want the neighborhood 'to *look* industrial, not *be* industrial'" (Zukin 1989: 104). But blame does not rest solely with the individuals who inhabit industrial zones and convert them into residential spaces; rather, as Winifred Curran argues, this process reflects the desires of the state: "The continued existence of manufacturing is seen as an impediment to the reconquest of downtown for high-end uses. Zoning, urban renewal, landmark status, tax breaks, and subsidies are all government policies that can encourage deindustrialization and create or constrain opportunities for gentrification in certain places" (2004: 1245–6). As formerly industrial spaces are displaced to make way for sites for artisanal production and craft consumption, industrial aesthetics stand in for actual industry.

This is certainly the case at Industry City, in Sunset Park, Brooklyn, a self-described 35-acre "innovation ecosystem."[6] The history touted on its website seamlessly connects the facility's origins in the 1890s as a "monumental intermodal manufacturing, warehousing and distribution center," through the 1960s, when urban manufacturing was in decline and Industry City suffered through a period of disinvestment and decay, to its revitalization beginning in 2013 through which it capitalized on the "rapidly emerging innovation economy" to once again become a thriving center of commerce and local employment.[7]

But as with the emergence of craft breweries discussed above, while Industry City harkens to its past, it reveals much more about the present and future as it blurs the boundaries between production and consumption while immersing visitors in a very gentrified Brooklyn. The "campus" directory divides its spaces into the following categories: creative and arts; design; government agency; media; nonprofit; office; production and manufacturing; retail; technology; and warehouse. Its "Makers Guild" features local and artisanal shopping and services (candles, art and ceramics, raw honey, and tattoos), and its food hall features the requisite combination of bars, breweries, and distilleries; coffee bars and bakeries; global cuisines (Middle Eastern, Japanese, Korean, Thai, Mexican); and local culinary artisans (cookies, chocolates, ice cream, breads, a vegan café, and an avocaderia).

At the same time, its aesthetics are familiar: millions of square feet of repurposed warehouse space featuring exposed pipes, brick walls, and concrete columns, corrugated metal siding, long wood tables for communal eating, repurposed lighting, and industrial signage. In short, Industry City is the quintessential gentrified Brooklyn space that harkens back to a long-lost industrial past while celebrating its contemporary approach to artisanal production and consumption. In Global Brooklyn, machines that were considered the driving force behind industrial production become controlled by people—for example, exposed steel and ductwork become scene-setting ornaments in spaces where knowledgeable baristas operate fancy coffee machines. In such instances, Colin Campbell's observations about the reappropriation of industry ring true:

> A certain kind of cultural capital is needed in order to envisage commodities as "raw materials" that can be employed in the construction of composite "aesthetic entities" and also to know what principles and values are relevant to the achievement of these larger constructions. In fact, craft consumers are likely to be people who do not merely possess just such cultural capital, but are also more concerned than most about the possible "alienating" and homogenizing effects of mass consumption—something that helps to account for their enthusiasm for the craft option, since they are likely to view this as the appropriate way of successfully resisting such pressures. (Campbell 2005: 35–6)

This embrace of the craft option is also seen as consumers seek out specialized knowledge and skills in the belief that by blurring the boundary between consumer and producer, they can be liberated from the industrial food system. For example, would-be students flock to specialized classes at The Brooklyn Kitchen, which recently relocated from Williamsburg to Industry City. A self-proclaimed "radical cooking school on a mission to change the world by teaching people how to cook like grown ups," their goal is to push back against a culture obsessed with "talking about, watching and photographing food." To do so, they invite students to focus on "where [food] comes from and how to prepare it."[8] However, this stance is ironic given that they rely heavily on social media—with its tantalizing images of perfectly marbled chops, and students triumphantly hoisting sausage garlands aloft—as an instrument for promotion. At the same time, their shuttered Williamsburg location remains available for film and photo shoots. Similarly, Fleishers Craft Butchery, with a location in the Park Slope neighborhood of Brooklyn, offers advanced classes aimed at "training the next generation of nose-to-tail butchers" as well as introductory classes for beginners (ages ten and up) and intermediate classes for experienced home cooks. As their website states, "In a time of increasing disconnection from our food systems," Fleischers' educational program seeks to "share the craft of butchery with our students in an effort to reconnect people to farms and food."[9] Such programs further demonstrate the desire of those who inhabit these neighborhoods to embrace artisanal production both in their purchases and in their own kitchens. This is part of a broader effort to create authentic identities that distinguish them from the masses who legitimately depend on the industrial food system without taking the ironic stance toward it that often accompanies hipster appropriations of industrially produced food.

The aesthetics and practices we describe in specific Brooklyn neighborhoods are linked to a much broader angst concerning international industrialization and global capitalism and the desire to affirm distinctive new identities and moralities. Recognizing this quest as a search for authenticity allows us to better understand Global Brooklyn's embrace of a specific ethos of consumption and production. Following Isabelle de Solier, we believe that "new moralities of self-making involve constructing what people believe to be a moral self through their taste in material objects" (2013: 15). Focused specifically on food choices and practices, de Solier (2013: 103) identifies many activities that are valued in the hipster enclaves of Brooklyn and their global counterparts, including eating local and in-season, and buying direct from producers at farmers' markets in an attempt to reinstate "face-to-face relations between producers and consumers."

De Solier (2013: 103) goes on to argue that "this process of re-embedding food is part of the reflexivity of late modernity, as some people return to what they see as traditional or preindustrial ways of life in their self-formation in response to the risks of postindustrial global modernity." Similarly, Campbell (2005: 37) focuses specifically on the hipster embrace of craft consumption, arguing that it has become "highly valued because it is regarded as an oasis of personal self-expression and authenticity in what is an ever-widening 'desert' of commodification and marketization." In these neighborhoods, authenticity is sought by embracing a set of practices that presumably stand in opposition to the disingenuous, conspicuous consumption embraced by modern capitalism.

As hipster food and drink spaces in Brooklyn seek to distinguish themselves from the mainstream through what they present as a countercultural ethos of consumption and production, it is important to recognize the extent to which they participate in, rather than resist, what Richard Florida identifies as new forms of capitalist enterprise that monetize creativity, innovation, and knowledge and integrate alternative and bohemian types into traditional economic institutions. According to Florida (2002: 57), businesses like those that are identified with Global Brooklyn represent the adaptability of capitalism as they "integrate formerly marginalized individuals and social groups into the value creation process." Similarly, Keiro Hattori, Sunmee Kim, and Takashi Machimura (2016: 173), in their work on Tokyo, point to the fundamental paradox of "global authenticity." Speaking of shoppers who are drawn to certain streets in Toyko, they write:

> They want to consume not just tangible things, but also an atmosphere of "nostalgia," "resilience," or even "resistance" to a widespread sense of economic uncertainty and cultural loss. They want to consume forms of intangible heritage such as local history and memory. [These] shopping streets survive because they manage to keep creating, more or less, the feeling of local authenticity that shoppers desire.

Of course, they point out the paradox that it is this very desire that has been commodified on these streets, where the quest for authenticity has itself become their "commercial 'niche.'" It is certainly possible to locate a similar paradox on the streets of Global Brooklyn, where desires to resist mainstream culture and to seek out a more authentic self have been commodified.

This paradox complicates any easy celebration of global expressions of the Brooklyn aesthetics, practices, and values identified throughout this volume. It also begs a larger question related to the impact of gentrification and the complexities of class politics: Who does and does not inhabit Global Brooklyn?

Before we fête "Global Brooklyn" neighborhoods as countercultural refuges from the harsh realities of modern capitalist hegemony and as a nostalgic return to a more authentic way of living and being, we need to consider the consequences of the spread of such neighborhoods across urban landscapes. Most significantly, we need to take into account the impact of gentrification on traditionally working-class, minority, and ethnic neighborhoods. As Elias le Grand reminds us, "Hipsters' appropriation of working-class spaces is part of their search for authenticity and 'edginess.' Yet the hipster has become a controversial social type, seen as contributing to both the trendiness of gentrifying neighborhoods and the displacement of working-class residents in the wake of rising rents and property prices" (2018: 185). Consequently, hipsters and their trendy neighborhoods have become associated with inequality and exclusion, thus bolstering the status quo.

For New York City, the consequences of this trend are significant, particularly as it extends deeper into Brooklyn neighborhoods. As Spencer Grover explains, "The boundaries of what is hip in Brooklyn are perpetually pushing outward. The same hipsters that fled Manhattan for Brooklyn 10 years ago are now skipping out of Williamsburg for Bushwick. And where young artists flock, the rest of the city seems to follow" (2017: np). As this happens, Brooklyn as both a literal and imagined space is changing. As Ben Adler (2014: np) writes, Brooklyn in this cultural moment becomes a shorthand for "hipster Brooklyn" or "gentrified Brooklyn," as if no other Brooklyn existed.

> Elite-oriented outlets are free to only cover Brooklyn's most affluent, Manhattan-adjacent neighborhoods: the artsy North Brooklyn of Williamsburg and Greenpoint and the leafy brownstone neighborhoods like Brooklyn Heights and Boerum Hill, though doing so leaves a host of compelling tales untold.

Similarly, Jana Kasperkevic notes that hipster Brooklyn is just a tiny (but overreported) fragment of a socioeconomically diverse borough, where a very large number of people live well below the national household income level and about 23% live in poverty (Kasperkevic 2014: np). These Other Brooklynites are increasingly being displaced by white, college-educated residents and as Philip Kasinitz and Sharon Zukin (2016: 31) note, "traditional, immigrant-owned mom-and-pop stores began to be replaced by art galleries, boutiques, and cafes: the global 'ABCs' of gentrification."

Important for understanding Global Brooklyn is recognizing that processes of gentrification extend from neighborhoods to popular culture itself. As Sarah Elsie Baker (2012: 634) explains, "The theory of aesthetic gentrification highlights the way in which individuals with high cultural capital are able to play with identity through the appropriation of old-fashioned or kitsch objects, while other people are marked as old-fashioned and tasteless." It seems to us that the elements of Global Brooklyn that go global are those that get "read" in spaces that enjoy high cultural capital, and that some of the very same elements may exist in other Brooklyn spaces but they do not get read as such. Even though they are also in Brooklyn, they are not sufficiently aesthetically gentrified to be part of Global Brooklyn. Anne Kadet, reporting for *The Wall Street Journal*, contemplates NYC neighborhoods where hipsters have yet feared to tread, including Bay Ridge, Canarsie, and Staten Island's South Shore, none of which saw an increase in average rents in the period between 2002 and 2014 (Kadet 2015: np). This is likely due to the fact that they gesture toward suburbia and they have relatively low walkability and transit scores, making them unpopular with millennial hipsters, who prefer communities with an urban feel (Kadet 2015: np).

The hipster embrace of urban spaces can also be read as a desire to claim an alternative route to authenticity, one that challenges the romantic assumption that only natural, pastoral settings offer true experiences unshaped by human desires. As Prentice (2001: 10) explains, "the proffering of authenticity frequently implies an anti-urban stance. Cities need therefore to find alternative means by which to proffer authenticity." It seems that the Global Brooklyn aesthetic steps in here to offer what feels like a more personal and authentic experience—the quirky noncorporate coffee shop, the quirky farm-to-table restaurant with the adorable tiny garden, the quirky bar with the venetian blinds, hodgepodge furniture, and full bocce court. In their search for authenticity, inhabitants of Global Brooklyn do not need to flee from cities; rather, they can reclaim urban spaces through a unique set of aesthetic choices and consumer practices that distinguish them from the mainstream culture they seek to resist.

In the end, it is clear that despite local differences, inhabitants of Global Brooklyn share a common desire to resist the global industrialization of late capitalism and the depersonalized commodification of late modernity. Suleiman Osman has described Brooklyn gentrification as "a saga of mixed intentions, sincere racial idealism mixed with disdain toward the nonwhite poor, and class populism blended with class snobbery" (Osman 2011: 16). We have examined the specific flows of capital (cultural and otherwise) that have characterized the emergence of the Global Brooklyn phenomenon, which we recognize as an expression of globalization. The embrace of Global Brooklyn results in distinction within the cultural context of a particular borough in New York City but also internationally, as, per Juliet Carpenter and Loretta Lees (1995: 288), its somewhat conforming symbols can be read by those in the know cross-nationally.

However, even as the other chapters in this volume provide a series of case studies on these global variations on a theme, we join them in cautioning against positioning Brooklyn as a point of origin for the values, aesthetics, and tastes attributed to it. Rather than identifying Brooklyn as the starting point for a global flow of knowledge, we would emphasize the porous nature of geographic and cultural boundaries, particularly in an age dominated by technology and social media. In this regard, we might consider the example of Peruvian hipsters discussed by Jace Clayton (2010), who suggests that their sudden embrace of previously looked-down-upon cumbia music followed the release of a compilation of cumbia music by a Brooklyn record label. This reveals a more dynamic relationship between Brooklyn and the global spaces to which it is connected. In this example, rather than exporting its music to Peru, Brooklyn's embrace of Peruvian music conferred a stamp of approval that functioned to recontextualize local Peruvian culture. We caution against any framework that may unintentionally reinforce a sense of US exceptionalism. Thus, this chapter has sought to recognize the extent to which Brooklyn neighborhoods and citizenry participate in a nuanced process of global exchange, rather than see the borough as the point of origin for the global phenomena explored in this volume.

Notes

1 https://www.covenhovennyc.com/
2 https://www.covenhovennyc.com
3 http://61local.com/our-philosophy/
4 http://61local.com/drink/
5 http://61local.com/our-philosophy/
6 https://industrycity.com/
7 https://industrycity.com/inside-ic/
8 https://www.thebrooklynkitchen.com/about/
9 https://www.fleishers.com/school/butcher-training/

References

Adler, B. (2014), "Stop using Brooklyn to mean hipster neighborhoods," *Columbia Journalism Review*, September 11. Available online: https://archives.cjr.org/behin d_the_news/brooklyn_reporting_elitism_hipster.php (accessed January 13, 2019).

Anderson, W., and DeSena, J. (1979), "The breweries of Brooklyn: An informal history." In R. S. Miller (ed.), *Brooklyn USA: The Fourth Largest City in America*, 125–36. New York: Columbia University Press.

Angst, M. (2018), "The 20 most hipster cities in the US—and why you should consider moving to one," *Business Insider*, April 10. Available online: https://www.thisisinsider.com/most-hipster-cities-in-america-movehub-2018-4 (accessed January 13, 2019).

Baker, S. E. (2012), "Retailing retro: Class, cultural capital and the material practices of the (Re)valuation of style," *European Journal of Cultural Studies*, 15 (5): 621–41.

Barthes, R. (1957), *Mythologies*. Paris: Editions du Seuil.

Bayley, S. (2017), "Brooklyn: Hipster hell or New York's greatest borough?," *The Telegraph* (UK), April 10. Available online: https://www.telegraph.co.uk/travel/destinations/north-america/united-states/new-york/articles/brooklyn-new-york-coolest-neighbourhood/ (accessed January 13, 2019).

Bridge, G. (2006), "It's not just a question of taste: Gentrification, the neighborhood, and cultural capital," *Environment and Planning A*, 38 (10): 1965–78.

Campbell, C. (1987), *The Romantic Ethic and the Spirit of Modern Consumerism*. Oxford: Blackwell.

Campbell, C. (2005), "The craft consumer: Culture, craft and consumption in a postmodern society," *Journal of Consumer Culture*, 5 (1): 23–42.

Carpenter, J., and Lees, L. (1995), "Gentrification in New York, London and Paris: An international comparison," *Journal of Urban and Regional Research*, 19 (2): 286–303.

Clayton, J. "Vampires of Lima." In Mark Greif, Kathleen Ross, and Dayna Tortorici (eds.), *What Was the Hipster? A Sociological Investigation*, 24–30. New York: n+1 Foundation, 2010.

Curran, W. (2004), "Gentrification and the nature of work: Exploring the links in Williamsburg, Brooklyn," *Environment and Planning A*, 36: 1243–58.

deSolier, I. (2013), *Food and the Self: Consumption, Production and Material Culture*. London: Bloomsbury.

Finn, S. M. (2017), *Discriminating Taste: How Class Anxiety Created the American Food Revolution*. New Brunswick: Rutgers University Press.

Florida, R. (2002), "Bohemia and economic geography," *Journal of Economic Geography*, 2: 55–71.

Gregor, A. (2015), "Crown heights, Brooklyn, where stoop life still thrives," *The New York Times*, June 17. Available online: https://www.nytimes.com/2015/06/21/realestate/crown-heights-brooklyn-where-stoop-life-still-thrives.html (accessed July 11, 2019).

Grover, S. (2017), "Coolest neighborhoods in NYC," *ELIKA Insider*, December 4. Available online: https://www.elikarealestate.com/blog/5-hippest-neighborhoods-new-york-city/ (accessed January 13, 2019).

Hattori, K., Kim, S., and Machimura, T. (2016), "Tokyo's 'Living' shopping streets: The paradox of globalized authenticity." In S. Zukin, P. Kasinitz, and X. Chen (eds.),

Global Cities, Local Streets: Everyday Diversity from New York to Shanghai, 170–94. New York: Routledge.

Kadet, A. "Where the hipsters haven't gone (Yet)." *The Wall Street Journal*, July 31, 2015. https://www.wsj.com/articles/where-the-hipsters-havent-gone-yet-1438375145

Kasinitz, P., and Zukin, S. (2016), "From 'Ghetto' to global: Two neighborhood shopping streets in New York City." In S. Zukin, P. Kasinitz, and X. Chen (eds.), *Global Cities, Local Streets: Everyday Diversity from New York to Shanghai*, 29–58. New York: Routledge.

Kasperkevic, J. (2014), "A tale of two Brooklyns: There's more to my borough than hipsters and coffee," *The Guardian* (UK), August 27. Available online: https://www.theguardian.com/money/us-money-blog/2014/aug/27/two-brooklyns-economy-hipsters-coffee (accessed January 13, 2019).

leGrand, E. (2018), "Representing the middle class 'Hipster': Emerging modes of distinction, generational oppositions and gentrification," *European Journal of Cultural Studies*, 23 (2): 184–200.

Linder, M., and Zacharias, L. S. (1999), *Of Cabbages and Kings County: Agriculture and the Formation of Modern Brooklyn*. Iowa City: University of Iowa Press.

Maly, I., and Varis, P. (2016), "The 21st-century hipster: On micro-populations in times of superdiversity," *European Journal of Cultural Studies*, 19 (6): 637–53.

O'Brien, F. (2018), "The hipster index: Brighton pips Portland to global top spot," *MoveHub*, April 19. Available online: https://www.movehub.com/blog/the-hipster-index/ (accessed January 13, 2019).

Ocejo, R. E. (2017), *Masters of Craft: Old Jobs in the New Urban Economy*. Princeton: Princeton University Press.

Osman, S. (2011), *The Invention of Brownstone Brooklyn: Gentrification and the Search for Authenticity in Postwar New York*. Oxford: Oxford University Press.

Prentice, R. (2001), "Experiential and cultural tourism: Museums and the marketing of the new romanticism of evoked authenticity," *Museum Management and Curatorship*, 19 (1): 5–26.

Ray, K. (2016), *The Ethnic Restaurateur*. London: Bloomsbury.

Scott, M. (2017), "'Hipster Capitalism' in the age of austerity? Polyani meets Bourdieu's new petite bourgeoisie," *Cultural Sociology*, 11 (1): 60–76.

Warf, B. (1990), "The reconstruction of social ecology and neighborhood change in Brooklyn," *Environment and Planning D: Society and Space*, 8: 73–96.

Zukin, S. (1989), *Loft Living: Culture and Capital in Urban Change*. New Brunswick: Rutgers University Press.

Dispatch

Chicago: Design of Displacement

Mireya Loza

I first started going to cafés in the mid-1990s as a student at the University of Illinois at Urbana-Champaign. These new spaces seemed a world away from my hometown of Chicago. At the coffee shop, I usually ordered an almond-flavored latte and a pastry and sat for hours, occasionally looking up and chatting when a friend would come by. I would come back home to my neighborhood and long for a coffee shop, not knowing exactly what I was asking for and what it meant. I could find many coffee shops in other areas of the city, but I wanted one within walking distance. By the time I made my way to graduate school in the early aughts, I came home one winter and noticed one five blocks away. Atomix seemed cool with a retro vibe; it had chairs and tables that resembled the ones I sat in during grade school. A giant mural of an astronaut (to fit with the space theme) adorned the longest wall. The giant colorful vibrant murals reigned supreme in Chicago, standing in contrast to the tiny thin lines that were used to compose the sleek minimalistic image.

I also suddenly found myself within a stone's throw from the hip, new brunch capital of the city, Wicker Park, where an eager restaurant market catering to the late-morning-to-late-lunch crowd had turned an old, tough multiethnic neighborhood into a hip mecca for brunching. Artists and baby-strolling yuppies descended upon the neighborhood and surrounding areas in search of real estate makeover opportunities and found willing sellers.[1] While most of those sellers were older, white ethnic property owners, many were also folks like my neighbors: Puerto Ricans and Mexicans. The incessant rise in property taxes and the difficulty of maintaining an aging home forced many working-class Latino families to sell away their properties right at the time when the neighborhood was "improving." And all those renters I grew up with seemed to have few options: they either lived in old apartment buildings that had not been remodeled in decades (much less ever), or they experienced being priced out of rehabbed shiny new apartments reserved for the new affluence moving into the neighborhood.

It seemed like every departure back to grad school put me through a time warp, and at every return, I felt like Rip Van Winkle waking and returning to my home to see it completely changed. My neighborhood was shifting at a disorienting pace, and I would ask myself: Have only four months passed or was it four years? Soon, I too had a cute cocktail bar two blocks from home, and one coffee shop led to another coffee shop, that led to a sushi bar that led to a brewery, that led to a pie shop, that led to a hipster market. Soon the affordable grocery store with bruised produce was replaced with the more expensive supermarket chain Dominick's that then gave way to an even more expensive Mariano's. An indispensable cultural institution for any Latinx neighborhood, the panaderias—small little bakeries filled with delicious sweet baked breads sold cheaply—were now closing down. It was as if the war on carbs had found a direct target and one by one the Mexican bakeries were taken out.

That same war on carbs seemed to miss the new pie shop Hoosier Mama, decorated in a quaint midwestern fashion and somehow standing as a flagpole that said to the new gentrifiers: "we are midwesterners, hear us roar" or, at least, "let us eat some damn good pie." I rode the wave, excited to impress worldly visitors from my new grad school life with a cocktail from a prohibition-era-inspired speakeasy. I too could appreciate the rediscovery of an ornate tin-plated ceiling or an old bank serving up its best charcuterie inside its vintage giant vault. But at every turn, there were fewer people who looked like me dining and many more serving. The new eateries were not only metaphorically but also literally catering to the new residents, and along with the panaderias, the taquerias began to close. A row of Mexican restaurants, some directly tied to a popular taqueria known as La Pasadita, and other knock-offs that had flourished in the 1990s, began to shutter. Somehow Mexican food in Mexican-owned restaurants, serving Latinx clientele, did not have the foodie vibe the new residents were yearning for and that opened up an opportunity for new hipster restaurants like Big Star and Antique Taco to reappropriate the taco in a new setting. Mexican taquerias in the neighborhood had missed the aesthetic hipster memo highlighting industrial, minimal, or kitsch.

The sit-down nice version of these traditional restaurants was most exemplified by Tecalitlán on Chicago Avenue, which was the spot that marked special occasions in my family. When my Tio Deciderio decided to return to Mexico in the 1980s, the whole family ate there to bid him farewell. The adults' plates were flanked by margarita glasses and Tecate bottles. When my mom graduated cosmetology school in the 1990s, we ate giant plates of enchiladas de mole and chile relleno, with sides of beans and rice. The big plate meant value and that no one would walk out hungry: working-class people had no interest in tiny plates. We would focus on full entrées of dishes that would take hours to prepare at home. The walls were lined with murals that served to remind the customers of Mexico, and you could see the artisanry in their heavy wooden chairs and tables that were made by hand. Business was going so well in the 1980s, groups of six or more dominated their waiting area.

Tecalitlán faced the 1990s with optimism, opening a new and more ambitious restaurant on Ashland Avenue, not too far from Chicago Avenue. This one would be grandiose, importing most of the ingredients from Mexico and taking its aesthetic inspiration from the haciendas in the state of Jalisco. But the seeds of gentrification were starting to sprout, and as rents rose and developers could sell suburbanites a new inner-city dream, their clientele base was slowly pushed out. Those who stayed on had less expendable income than they had before. This change created an opening for Mexican restaurants owned by people of the right hue, who had access to large amounts of capital and could "reinterpret" Mexican food for a different client. Their spaces would not look like traditional gringo-owned Mexican restaurants, big sombreros and *papel picado* adoring the interiors: the new spaces would be modern with a hint of industrial, and tacos instead of full plates would reign supreme. There would be no sides of rice and beans.

These new Mexican restaurants could capitalize on the labor of the Mexican workforce so that those who peeked into the kitchens could still see authentically brown bodies preparing their tacos. Commonly served drinks like Micheladas and Palomas could be discovered by these new customers, who did not have to be reminded of a Mexico left behind and could build on aesthetic fantasies of an imagined Mexico or discard that all together. Restaurateurs could learn from the people who had been cooking in restaurants all along and could turn this knowledge into dollars and cents by repackaging it. They could also amplify the message by investing in an interior aesthetic that was circulating on Instagram, Facebook, and Twitter. To be sure, this aesthetic was not cheap and often required significant capital. In a city that was home to the second-largest urban concentration of Mexicans in the United States, gentrification hit neighborhoods but also plates.

While new aesthetics of Mexican food were shaping my neighborhood's foodscapes, restaurants in the historically Mexican neighborhood of Pilsen were also feeling the squeeze. Chefs like Alfonso Sotelo had to pull on a new toolkit that went beyond taste to cater to both new and old clientele. While he did not have access to the capital that created the new vanguard of hipster restaurants, he held a deep understanding of the aesthetics of a plate. Born in Guerrero, Mexico, Chef Sotelo made his way to California in the early 1990s and eventually settled in Pilsen in 1996 to join his family residing in the neighborhood. For nineteen years he worked in Rick Bayless's restaurants Topolobampo and XOCO, learning what Bayless excelled in: how to present Anglos with the modern Mexican food for which they yearned. Rick Bayless made a name for himself as the chef who mastered Mexican cuisine, creating a series of restaurants that would garner the attention of gourmands.

Chef Sotelo worked in an environment that catered to a high-end clientele in search of the best Mexican food the city had to offer, and after his long tenure there he set his sights on opening his own restaurant. He combined food memories of his grandmother's cooking along with his work experience with Bayless, and the result was 5 Rabanitos, which opened in 2016. He focused on a restaurant space that had previously been home to a pupuseria, right on 18th Street, the main commercial artery of Pilsen. He understood the geographic potential of the space, as it was catty-corner from Harrison Park, home of the National Museum of Mexican Art. When he opened, he knew that many of the changes that would have to take place would be carried out slowly, since he did not have the capital to gut or fully remodel the space. He told himself: "I would have to make do with whatever was inside." "When we opened, we started with table cloths covered with plastic because we didn't have enough money," he described. "Little by little, we changed things, modified, and bought another style of tables, and brought in other chairs and lots of dishes."

He shared that when they first opened, they did not have enough pots, pans, and dishes. But what he lacked in financial capital, he made up with one delicious dish at a time at an affordable price point. His food would not be out of reach for the longtime residents of Pilsen. His approach to the plate was made up of "design, taste, and style." Unlike many of the new darlings in the food scene, the décor was DIY: "Nothing was bought . . . some of the frames I brought from home because I needed to design the restaurant with things that I could not buy," and some of the art was gifted to him by clients. And while Chef Sotelo has received a wave of positive attention and reviews for his restaurant, he is also keenly aware that the entire neighborhood is undergoing massive changes because of gentrification.

Down the street, restaurateurs with access to large amounts of capital bought a Bohemian hall built in 1892 and refashioned this turn of last century building into three spaces: a music venue, a restaurant, and a bar. Their Michelin-starred restaurant Dusek's Board and Beer serves up food to the hipster clientele that want to eat in a structure that builds on their ethnic European fantasies rooted in reclaiming not only the space but the entire neighborhood, arguing that it was Bohemian long before it was Mexican. The views of the past in these gentrified spaces are as much about class as they are about race. The newcomers conveniently forgot that many of their grandparents abandoned the city center generations ago and that Mexican immigrants took dilapidated housing stock, storefronts, and buildings and remade main commercial thoroughfares in the city.[2]

Chef Sotelo's restaurant stands at the crossroads of Global Brooklyn, attempting to mediate the designs of plates and spaces, while also making larger neighborhood claims. "I want our Hispanics, our culture, not to leave Pilsen. . . . [I want them] to try to hold on as long as they can and not sell their homes . . . so that Pilsen can be like it has always been and always home to our Mexican culture." He understands that in the face of gentrification many longtime homeowners are finding it hard to stay in the neighborhood, but he hopes Pilsen's Mexican community can hold on. While we do not know what will happen to Pilsen, we do know that despite the economic challenges, Chef Sotelo is carving out a space of his own, his beautifully designed plates posted on Facebook, Twitter, and Instagram. He adopted the aesthetic lexicon by setting food on large white plates versus the more folksy version of beige or brown serving dishes or even the bright look of the ever so popular fiestaware. Chef Sotelo arranges his dishes in ways that avoid the schema of rice and beans on the sides of large plates but fashions them in a style that is appealing to a downtown crowd. For longtime residents, Chef Sotelo and a new cadre of Latinx chefs in Pilsen represent the hope that the Mexican community, Mexican-owned restaurants, and ultimately Mexican food will resist gentrification and erasure in the culinary landscape of the neighborhood shaped by Global Brooklyn.

Notes

1 Richard Lloyd's ethnographic study provides a snapshot of these neighborhoods in the midst of 1990s gentrification (Lloyd 2005).
2 For more on white flight and urban renewal in Pilsen, see Amezcua 2017.

References

Amezcua, M. (2017), "Beautiful urbanism: Gender, landscape, and contestation in Latino Chicago's Age of Urban Renewal," *Journal of American History*, 104 (1): 97–119.

Lloyd, R. (2005), *Neo-Bohemia: Art and Commerce in the Postindustrial City*. London: Routledge.

Conclusion

Thinking Food through Design

Fabio Parasecoli and Mateusz Halawa

Exploring different manifestations of Global Brooklyn around the world, the chapters and the dispatches in this volume have pointed to the growing complexity of the cultural formation and the design style that supports it. Rather than providing definitive answers to the provocations we offered in the Introduction, the authors' contributions have raised further questions, showing how Global Brooklyn is a moving target both in terms of its meaning and its dispersion dynamics. In this concluding chapter, we reflect on some of the themes that have emerged through the global wanderings of the authors in this volume, while highlighting areas and themes that deserve further research and analysis.

The worldwide circulation of Global Brooklyn requires a reflection on its connection to space and geography. Although visible in most large cities, anxious gentrifiers and hipsters in denial—with their fetishization of working-class blue-collar manual activities into consumable experiences—tend to be particularly active in specific and highly recognizable areas in the urban landscape. In fact, Global Brooklyn establishments often open in each other's vicinity, creating small enclaves where cafés, organic food markets, artisanal chocolateries, and natural wines stores are found next door to independent bookstores, craft and stationary shops, and bike repair workshops. This generates urban ecologies that are often celebrated as seeds of new forms of inhabiting cities and creating social connections, all while constituting leisure districts that, in reality, are financially accessible and culturally legible to relatively small segments of the population. Customers patronizing these stores can experience a sense of place that reinforces their ethical and aesthetic choices and ideals, reflected not only in the objects and in the interior design but also in the urban space and in the interactions among those who shop or eat there.

These enclaves do not appear and grow in isolation. They often create ties with similar locations in other cities based on shared experiences, common purveyors of ingredients, materials, and products, as well as the physical circulation of both producers and consumers. Furthermore, each shop, café, and restaurant is also connected with far-flung outposts such as cocoa and coffee plantations, natural wineries, and organic farms, reinforcing the feeling—and the pleasure—of participating in an alternative economic network. Unlike previous modes of placemaking under globalization, theorized by Marc Augé (2009) as productive of non-places, Global Brooklyn's circulation articulates globality but also, at the same time, a specific kind of being in the world that is built around the experience of locality: the global and the local constitute each other through connections that trouble any simplistic analysis of the relationship between center and periphery (Appadurai 1996; Wilk 2006). Still, this process is far from being entirely horizontal and democratic. As not all locations enjoy the same political and economic clout on the global scene, their connections may actually repeat the complex relationships of power embedded in the neoliberal economic structures imposed since the 1980s by the Washington consensus, the establishment of the World Trade Organization, and the financialization of food commodities (Harvey 2005; Parasecoli 2019).

Cities are the nodes where complex global networks of political and financial power and transnational flows of people, capital, objects, and information interact with actual territories (Sassen 2001). Many of the Global Brooklyn establishments operate, in fact, in what scholars are calling superstar cities, agglomerations that contribute to the production of durable inequalities (Le Galès and Pierson 2019). As Doreen Massey reminds us, world cities are not necessarily ideal places brimming with opportunities that tend to diminish inequalities, but rather inevitable destinations for often brutal migration processes set in motion by cities themselves and by the powers they congeal (Massey 2007). While owners and front-of-the-house staff are often educated, upwardly mobile urban dwellers, some of the less visible workers in Global Brooklyn establishments may well be undocumented immigrants or may belong to underprivileged segments of society.

While contributors in this volume have focused on consumer spaces or urban experiences where production is closely connected—also geographically—to consumption, from upscale stores to market halls, we cannot forget that the cultural formation has extended beyond city limits. It is possible to recognize the presence of Global Brooklyn elements in culinary tourism, from wineries to farms, where the revaluation of manual skills is particularly strong, although still surrounded by carefully designed environments. It is also visible in new modes of representation of the rural on Instagram and other platforms, often by actors whose lifestyle is deliberately bridging the urbane cool with rural "authenticity." These are rural locations with an urban flair, arranged to respond to the preferences and practices of urban consumers.

Global Brooklyn is a dynamic of refraction that brings together the provincial with the cosmopolitan: the aesthetic formation may find expression in Lublin via Warsaw via Berlin via Brooklyn via Kyoto. Each node is an active part of various layers of communication: global, regional, national, and local, with different elements making sense in different layers. Individuals and communities of producers, entrepreneurs, chefs, and consumers operate in ways that, despite their embeddedness in specific places, constantly move beyond them only to re-embed themselves, deliberately and reflexively. This modality is a product of a society of networked individuals that through social media are able to establish connections with far-away people, sometimes more easily than with neighbors, creating global tribes that share language, imaginary, a visual landscape and, as a consequence, an aesthetic regime and a common repertoire of practices and materialized values.

The mode of dispersion of Global Brooklyn that we described as "decentralized sameness" may as well manifest itself also in other domains and aspects of globalization. The metaphor of the rhizome, as elaborated by Deleuze and Guattari (1987), may provide a productive conceptual framework. A rhizome, Global Brooklyn is not like a tree with a single root, so genealogy and tracing descent may not be the right approaches to understand its diffusion. It is rather a multiplicity that assumes very diverse forms, emphasizing connections and heterogeneity, while revealing the inherent multiplicity and apparent disorder of any cultural formation. Each of its nodes can connect with any other node, with no preestablished order or fixed dynamics. Rhizomes expand across different surfaces, connecting elements into a temporary arrangement, teeming with inordinate life. Similarly, Global Brooklyn manifestations originate, expand, and evolve not in abstract spaces, but in the process of making actual places: food practices and materialities allow languages, bodies, politics, economics, sciences, and other fields of experience to connect in unpredictable fashion through deterritorialization and reterritorialization. By doing so, the whole of the Global Brooklyn aesthetic regime inevitably opens itself to change, because the expansion always mutates in terms of its internal dynamics and its characteristics.

Suggestions for ways to interpret the Global Brooklyn mode of dispersion can also be found in Zygmunt Bauman's (2000) reflections about liquid modernity, in which identity, relationships, and even economic connections are in constant shift, both in space and in time. In *Le vespe di Panama* (The Wasps of Panama, 2007), a small book he published in Italian only, Bauman writes about "swarms," arguing that the "centrality of the center" has been dismantled and "the ties among closely connected and coordinated spheres have been broken" (19, our translation). As a consequence, it is necessary to rethink tensions such as center/periphery and superiority/inferiority, without asserting that humanity is heading toward greater equality. Bauman argues that we now live in the epoch of networks that look for their roots not in the past but rather in the present and are focused on individuals (23). Like in the case of Global Brooklyn, such networks are extremely flexible, and their content and composition can easily be modified. "Within networks, 'belonging' does not precede, but rather follows identities. Belonging to a network, compared to belonging to a traditional totality, such as a group, tends to become an extension of shifting identities, quickly and meekly following the successive renegotiations and redefinitions of identity" (25). The participants in Global Brooklyn appear to be behaving like a swarm, in which every unit "repeats the moves made by any other units, executing the task in its entirety, from start to end and in all its parts, individually (in the case of swarms of consumers, the executed task is consumption)" (28). This would support an interpretation of the Global Brooklyn "decentralizeddistributed sameness" as a form of dynamic, collective movement of independent actors that nevertheless determine their trajectory in relation to each other. However, unlike in Bauman's swarms, not every unit is equally positioned in terms of cultural and economic power; moreover, leaders and hierarchy emerge in terms of global prestige and visibility.

Imagination, examined by Appadurai as a "collective, social fact" and as "part of the quotidian mental work of ordinary people" (1996: 5), also seems to play some role in the dispersion of the Global Brooklyn experience. Imagination is "neither purely emancipatory nor entirely disciplined but is a space of contestation in which individuals and groups seek to annex the global into their own practices of the modern" (4). Puritanical discounting of forms of imagination such as those expressed in Global Brooklyn establishments needs to be avoided, because "where there is consumption there is pleasure, and where there is pleasure there is agency" (7). Appadurai's analysis of global cultural flows reminds us that those dimensions "are not objectively given relations that look the same from every angle of vision but, rather, . . . they are deeply perspectival constructs" (33). Despite its apparent internal similarities, Global Brooklyn and the imaginary that underpins it look and feel differently, depending on where its actors are located.

The understanding of the Global Brooklyn dispersion model through the lenses provided by Deleuze and Guattari, Bauman, and Appadurai, as well as the concept of "decentralized sameness" we propose in this volume, contributes to broader discussions within food studies in its focus on globalization and worldwide diffusions of cultural formations. Such reflections question on the one hand approaches that embrace a democratic and egalitarian outlook on food-related dynamics and on the other hand perspectives that instead focus on hierarchical structures and tensions such as local/global and center/periphery.

In this volume, chapters and dispatches also took up the relationship of Global Brooklyn with social class. As we developed the project with collaborators, their contributions pointed to the centrality of the logics of gentrification and urban renewal, transformation of professions, elevation of craft, and the significance of consumption in notions of respectability, worldliness, and the good life. However, in each locale, these problems are inflected by particular histories or conflicts: in Warsaw, they may speak to anxieties around Europeanization and unequally distributed benefits from the economic boom of the past decades; in Mumbai, they may materialize a conflicted relationship between nomadic elites and the disenfranchised local populations, while in Chicago they may take on the form of commodification of ethnicity. In large cities, Global Brooklyn has established a complex relationship with foodie cultures and the so-called food movements, which all have priorities, practices, and forms of communication that only partly overlap. A sensitive aspect of these interactions is the phenomenon of food gentrification (Mikki Kendall's term, see Staley 2018), indicating the displacement of affordable sources of food to make space for more upscale and varied, better designed, but more expensive establishments that reflect the preferences of the newcomers.

Many contributors were less concerned with creating a stable and neat stratification, and more with revealing the degree to which matters of design around food and drink dynamically shape the lived experience of class (Thompson 1963). Especially in places like Warsaw or Mumbai, Global Brooklyn is less about passively expressing a given social status, and more about *reaching* for it, performatively trying to become someone or some place. The classed dynamics of Global Brooklyn will lean toward the aspirational processes that Chipkin (2012) has called *middle classing*: "efforts being made and a journey undertaken rather than an arrival point or a concrete category" (James 2017: 2). Another point of interest is raised by the marginal figures of hipsters and craftsmen whose everyday practice plays on the boundaries between white collar and blue collar, tests the shifting grounds of respectability and status, and improvises new classed imaginaries. Lest the figure of the hipster lead us astray, we have shown that the phenomenon cuts deeper than fashion. The emerging aesthetic regime aligns with powerful socioeconomic processes ranging from postindustrialization and gentrification of cities, to struggles around justice and sustainability in the food system, as well as the rise of the digital.

In Global Brooklyn, the social cannot be separated from the material. Its important component that several contributors to this volume have explored is the very nature of aesthetic regimes, which the anthropologist Krisztina Fehérváry defined as "politically charged assemblages of material qualities that have provoked widely shared affective responses" (Fehérváry 2013: 3). Studying the built environment left from socialist Hungary, she suggests that the material qualities of things, or *qualia*, "came to provoke affective responses to the sociopolitical and economic ideologies with which they were aligned" (Fehérváry 2013: 3). It can be argued that Global Brooklyn similarly creates affective responses, not only through the qualities of interiors and objects but also in the way such materialities get entangled in the relations between food providers, consumers, and their ethical positions. Political scientist Davide Panagia (2009) has explored how sensations, which "register on our bodies without determining a body's nature or residing in any one organ of perception" (2), do influence how individuals and communities engage in political life. He goes as far as to propose "a new political subjectivity: the tasting subject" (3), who is open to interruptions, dissent, and interactions with others. In the context of Global Brooklyn the sensorium is made not only of building materials and spaces but also of flavors and scents. However, the same qualia may play differently in different contexts. Examining Global Brooklyn locations, it is easy to observe that the approach is adopted across various modalities, with intensity and completeness on a spectrum that goes from the wholesale acquisition of the cultural formation (materials, practices, symbols, and discourses) to the mere use of selected elements of the sensorium, used as cool design references and hints to interior design trends.

These processes find their material and sensory arenas and create lasting effects in Global Brooklyn. Retrofitted reclaimed wood and iron in the interiors, online platforms of visual communication, a claim to a renewed aesthetics and ethics of eating, drinking, and service labor all point to efforts at establishing and transforming values around food by way of manipulating their materiality (Fehérváry 2013). From this perspective, this research has sought to enhance the methodological and theoretical toolkit of food studies, which at times does not engage enough with the materialities of objects and constructed environments surrounding food. The field tends to focus instead on food itself and its sensory traits in their relationship with embodied experiences, memory, identity, culture, and the social relations built around them (Julier 2013; Sutton 2001). The study of sensoria from a design point of view and a greater attention to the impact that tools and technologies have on the way we relate to food can provide an effective lens to observe familiar activities such as cooking, brewing, and baking, further eroding the distinction between subject and object, spirit and matter, that has already been the target of critique in food studies (Curtin and Heldke 1992; Heuts and Mol 2013; Perullo 2016). Expanding the conversation between food studies and the burgeoning field of food design can generate a better understanding of the complex relationships between materialities and affect, between the utility of objects and their meaning in practice (Horowitz and Singley 2004; Vodeb 2017; Wrigley and Ramsey 2016).

Global Brooklyn reminds us that the social and political relevance of materiality and built environments cannot be discounted. As some of the contributors to this volume have pointed out, designed environments can be conducive to forms of conviviality that aim to provide joyous and fulfilling experiences, transcending food and its production or consumption to provide opportunities to envision and experience what different futures may feel like. The sensory qualities of objects and the distribution and flow of spaces in which we live and operate reflect and at times reinforce a wide array of moral and ethical projects. They range from the profit-oriented use of design elements and the exploitation of global trends to the desire to counteract the most unsustainable and unjust aspects of the industrialized food system, through DIY practices and the elaboration of visions of care and community building. However, any attempt at resistance and subversion still takes place within a capitalist framework. Furthermore, Global Brooklyn actors are at times not particularly self-reflective in terms of who is inherently excluded from their projects because of education, class, or ethnicity, to mention but a few determining factors.

The unrecognized—and at times willfully ignored—social tensions that underpin the expansion of Global Brooklyn raise the legitimate question whether this cultural formation is just another shallow, passing trend, or something deeper. These reflections are closely connected with one of the underlying challenges that has accompanied the emergence of food studies as an academic field of research and teaching that spans from systemic to cultural issues, from quantitative to qualitative approaches, from practical interventions to purely analytical reflections. Studying ever fleeting and emergent food-related cultural formations may seem to some vapid and irrelevant, but in fact, it can provide an entryway to understand social, economic, and political tensions underlying and shaping the global food system, and not exclusively in terms of consumer culture. The very fact that not only transnational corporations, from fast food to hotels, but also companies coming out of the Global Brooklyn experience itself are trying to capitalize on the aesthetic regime in terms of look and feel, points to the relevance of the phenomenon as a focal point of real consumer concerns and business investments. This volume has offered methodological and substantial suggestions about how to look at dynamics of corporatization and entanglement in neoliberal processes of a food-related cultural formation, and how an ethical project—if there is one—gets emptied out and repurposed. By providing support, training, and education around the world, companies like the Department of Brewology (operating in the beer sector) promote a Global Brooklyn approach, but at the same time, they turn it into a marketable constellation of commodities and paint-by-numbers services. Is their modus operandi a contributing factor of the "decentralized sameness" we have proposed in this volume? Are they just central cultural and market intermediaries responding to pull factors, or are they pushing specific visions and approaches as part of their business model?

New cultural formations may have been emerging in parallel with Global Brooklyn. For all we know, a new trend in restaurant and café culture may already be out there, still undiscovered, or at least unrecognized as a global phenomenon. We hope that this Global Brooklyn project has generated effective theoretical perspectives and a methodological toolkit bridging food studies, ethnography, and design for a more critical and creative exploration of food futures.

References

Appadurai, A. (1996), *Modernity at Large: Cultural Dimensions of Globalization.* Minneapolis: University of Minnesota Press.

Augé, M. (2009), *Non-Places: An Introduction to Supermodernity.* New York: Verso.

Bauman, Z. (2000), *Liquid Modernity.* Cambridge: Polity Press.

Bauman, Z. (2007), *Le vespe di Panama: Una riflessione su centro e periferia.* Bari: Laterza.

Chipkin, I. (2012), *Middle Classing in Roodepoort: Capitalism and Social Change in South Africa,* Public Affairs Research Institute, Long Essays. Vol. 2. Johannesburg.

Curtin, D. W., and Heldke, L. M. (1992), *Cooking, Eating, Thinking: Transformative Philosophies of Food.* Bloomington: Indiana University Press.

Deleuze, G., and Guattari, F. (1987), *A Thousand Plateaus: Capitalism and Schizofrenia*. Minneapolis: University of Minnesota Press.

Fehérváry, K. (2013), *Politics in Color and Concrete: Socialist Materialities and the Middle Class in Hungary*. Bloomington: Indiana University Press.

Harvey, D. (2005), *A Brief History of Neoliberalism*. New York: Oxford University Press.

Heuts, F., and Mol, A. (2013), "What is a good tomato? A case of valuing in practice," *Valuation Studies*, 1 (2): 125–46.

Horowitz, J., and Singley, P. (2004), *Eating Architecture*. Cambridge, MA: MIT Press.

James, D. (2017), "Not marrying in South Africa: Consumption, aspiration and the new middle class," *Anthropology Southern Africa*, 40 (1): 1–14.

Julier, A. (2013), *Eating Together: Food, Friendship, and Inequality*. Urbana: University of Illinois Press.

Le Galès, P., and Pierson, P. (2019), "'Superstar cities' & the generation of durable inequality," *Daedalus*, 148 (3): 46–72.

Massey, D. (2007), *World City*. Malden, MA: Polity Press.

Panagia, D. (2009), *The Political Life of Sensation*. Durham and London: Duke University Press.

Parasecoli, F. (2019), *Food*. Cambridge, MA: The MIT Press.

Perullo, N. (2016), *Taste as Experience: The Philosophy and Aesthetics of Food*. New York: Columbia University Press.

Sassen, S. (2001), *Global City: New York, London, Tokyo*. Princeton: Princeton University Press.

Staley, W. (2018), "When 'Gentrification' Isn't About Housing," *The New York Times*, June 8. Available online: https://www.nytimes.com/2018/01/23/magazine/when-g entrification-isnt-about-housing.html

Sutton, D. (2001), *Remembrance of Repasts: An Anthropology of Food and Memory*. New York: Berg.

Thompson, E. P. (1963), *The Making of the English Working Class*. London: Gollancz.

Vodeb, O. (2017), *Food Democracy: Critical Lessons in Food, Communication, Design and Art*. Chicago: Intellect.

Wilk, R. (2006), *Home Cooking in the Global Village*. Oxford: Berg.

Wrigley, C., and Ramsey, R. (2016), "Emotional food design: From designing food products to designing food systems," *International Journal of Food Design*, 1 (1): 11–28. doi:10.1386/ijfd.1.1.11_1

Notes on Contributors

JT Akai is a published Ghanaian author and writer. His writing interests include afro-glamor and Afro-fantasy. He has published on both traditional and urban Ghanaian folklore. His most recent books are *Madam High Heel* and *Afi, So Special*, which are available on Amazon.

Alexandra Forest is a graduate student at California State University, Fullerton. There she is studying visual rhetoric and its intersection with critical culture.

Dr. **Helena Grinshpun** is currently an adjunct lecturer at the Asian Studies Department, the Hebrew University of Jerusalem, and the Asia Unit Coordinator at the Harry S. Truman Research Institute for the Advancement of Peace. She carries a PhD in anthropology and Japanese studies from Kyoto University, Japan. Since 2010, she has been teaching courses on Japanese contemporary society and culture at the Hebrew University's Asian Studies Department. Her main research interests are urban and material culture, structuring of public and private space, consumer behavior, community and family in Japan. She has published several articles on global coffee and consumption in Japan; her book manuscript on coffee culture and public space in Japan is to be published by Routledge.

Professor **Liora Gvion** is a qualitative sociologist and senior lecturer who teaches at the Kibbutzim College of Education and lives and works in Tel Aviv, Israel. Her areas of expertise are the sociology of food and the sociology of the body. Her research revolves around the social and political aspects embedded in Palestinian food in Israel, the anorectic body, the operatic body, and the ethnic lesbian body. She completed a big study on MasterChef Israel in which she identified national and ethnic themes that underlie the show.

Mateusz Halawa is a researcher at the Max Planck Partner Group for the Sociology of Economic Life at the Institute of Philosophy and Sociology of the Polish Academy of Sciences in Warsaw and a doctoral candidate at the Department of Anthropology of the New School for Social Research in New York. He focuses on the relationships between finance, design, and social transformations. He has published in *Cultural Studies, Journal of Cultural Economy*, and *Food, Culture and Society*.

Priyansha Jain is a curator and designer. Her interest in food as an artistic medium often manifests into experiential assemblies to reflect about the past in the present to create an alternate future. Currently, she is Deputy Director at Mumbai Art Room, a curatorial lab in Bombay that creates opportunities for young curators in India and the world. She is also part of a Berlin-based collective called empty_glass. In 2017, she co-founded studio_exforma to realize interdisciplinary and conceptual projects in India.

Kathleen LeBesco, PhD, is Associate Vice President for Strategic Initiatives and Professor of Communication and Media Arts at Marymount Manhattan College in New York City. Her work concerns food and ideology, fat activism, disability and representation, working-class identity, and queer politics. She is author of *Revolting Bodies: The Struggle to Redefine Fat Identity*, coauthor of *Culinary Capital*, and coeditor of *The Bloomsbury Handbook of Food and Popular Culture, Bodies Out of Bounds: Fatness and Transgression, Edible Ideologies: Representing Food and Meaning*, and *The Drag King Anthology*, as well as dozens of book chapters and journal articles. Her work was nominated for a Lambda Literary Award and received an honorable mention for the Sylvia Rivera Award in Transgender Studies. She has appeared under the pseudonym GB as a commentator on *The Today Show, National Public Radio*, and *Les Francs-Tireurs* (CBC), in documentary films, and in print outlets including *The New York Times, The Chronicle of Higher Education*, and *Der Spiegel* (Germany).

Jonatan Leer is Head of Food and Tourism Research at University College Absalon, Roskilde. He has published widely on food culture, notably on the new Nordic cuisine, food pedagogies, meat consumption and the gendering of food practices, in journals such as *Food, Culture and Society, Food and Foodways, European Journal of Cultural Studies*, and *Feminist Review*. He has written three books in Danish and edited the book *Food and Media* (2016) and contributed to *Food and Age in Europe* (2019), *Alternative Food Politics* (2018), *Food and Popular Culture* (2017), and *Food, Masculinities, Home* (2017). Jonatan is a visiting lecturer at the University of Gastronomic Sciences in Pollenzo, Italy, and a member of the Danish gastronomic academy.

Tania Lewis is the codirector of the Digital Ethnography Research Centre and Professor in the School of Media and Communication at RMIT University,

Melbourne, Australia. Her research critically engages with the politics of lifestyle, sustainability and consumption, and with global media and digital cultures. Tania has published over fifty journal articles and chapters, four edited collections and has authored and coauthored four monographs: *Digital Food: From Paddock to Platform*, *Smart Living: Lifestyle Media and Popular Expertise*, *Telemodernities: Television and Transforming Lives in Asia*, and *Digital Ethnography: Principles and Practices*.

Mireya Loza is an assistant professor in the Department of History at Georgetown University. She earned her doctorate in American Studies and an MA in Public Humanities at Brown University. Her areas of research include Latinx History, Public History, Labor History and Food Studies. Her book, *Defiant Braceros: How Migrant Workers Fought for Racial, Sexual and Political Freedom*, examines the Bracero Program and how guest workers negotiated the intricacies of indigeneity, intimacy, and transnational organizing. Her first book won the 2017 Theodore Saloutos Book Prize awarded by the Immigration and Ethnic History Society and the Smithsonian Secretary's Research Prize. Her research has been funded by the Ford Foundation, the National Endowment for the Humanities, and the Smithsonian's Latino Center.

Bryan W. Moe is an assistant professor at Biola University located in Southern California. He teaches in the Communication Studies Department and specializes in rhetoric. His research has been mostly focused on social movements and food, particularly the food truck and street food movements post-2008 US recession.

Peter Naccarato is Interim Vice President for Academic Affairs and Dean of the Faculty at Marymount Manhattan College. With Katie LeBesco, he is coauthor of *Culinary Capital* (2012) and coeditor of *The Bloomsbury Handbook of Food and Popular Culture* (2018) and *Edible Ideologies: Representing Food and Meaning* (2008). With Zach Nowak and Elgin Eckert he is coeditor of *Representing Italy through Food* (2017). He has developed and taught interdisciplinary courses in food studies at Marymount and at the Umbra Institute, an American study abroad school in Perugia, Italy. He also serves on the advisory board of the Institute's Center for Food & Sustainability Studies.

Thiago Gomide Nasser is a founder and organizer of Junta Local, a platform that brings together small food producers and consumers in Rio de Janeiro, Brazil. He earned his PhD in Political Science from the Instituto de Estudos Sociais e Políticos of the State University of Rio de Janeiro researching the role of chefs and the food media in framing political and social issues related to the food system. He is also recently helped launch *Revista Feira*, a food magazine based in Rio.

Yoshimi Osawa is an assistant professor at Aoyama Gakuin University, Japan. She specializes in the anthropology of food and ethnobotany, and her research centers on understanding of relationship between humans and nature, particularly by looking at food, ecology, and human sensory perceptions. Her current research projects include the history of MSG (monosodium glutamate) consumption and its rejection as well as traditional food practice in Thailand..

Fabio Parasecoli is Professor of Food Studies in the Nutrition and Food Studies Department at New York University. His current research explores food in cultural politics, particularly in heritage and design. Recent books include *Feasting Our Eyes: Food, Film, and Cultural Citizenship in the US* (2016, authored with Laura Lindenfeld), *Knowing Where It Comes From: Labeling Traditional Foods to Compete in a Global Market* (2017) and *Food* (2019).

Adriana Rosati is a designer with background that includes graphic design, corporate identity, photography, and a career as chef and caterer. She is a trained graphic designer, with a MA in Design Studies at Central Saint Martin's in London. Currently she is Senior Graphic Designer at Pell Frischmann Consulting, coediting the Asian Cinema website *Asian Movie Pulse*, and writing as UK correspondent for the cinema website *Link in Movies*.

Dr. **Signe Rousseau** is a lecturer in critical literacy and professional communication at the University of Cape Town, where her doctoral and postdoctoral research focused on the phenomenon of celebrity chefs and the increased politicization of and attention to a profession which used to remain largely unseen. She is the author of *Food Media: Celebrity Chefs and the Politics of Everyday Interference* (2012), and *Food and Social Media: You Are What You Tweet* (2012), a contributing author to a number of edited volumes and reference works on various aspects of food and media, and a cochair of the Editorial Collective of *Gastronomica: The Journal for Food Studies*.

Dr. **Susan Taylor-Leduc** earned both her masters and doctoral degrees from the University of Pennsylvania. Since 1992, she has worked as a teacher, curator, and university administrator in Paris. She is currently completing a book entitled *Designing Legacy: The Picturesque Garden in France 1775–1867* for Amsterdam

University Press and is affiliated with the Centre des Recherche du Château de Versailles. She is the founder and director of Picturesque Voyages, a cultural travel company.

Oliver Vodeb is senior lecturer at RMIT University in Melbourne, where he teaches and researches in the Master of communication design. He is principal curator of Memefest festival of disobedient communication and design and of Lipstick+Bread, cooking classes as pleasure, conversation, and culture. His last book is *Food Democracy*, published by Intellect Books, UK. He is currently working on a new book project titled *Radical Intimacies*.

Index